Introductory Biogeography

Introductory Biogeography

Edited by
Emma Darling

▤ Larsen & Keller
www.larsen-keller.com

Introductory Biogeography
Edited by Emma Darling
ISBN: 978-1-63549-696-3 (Hardback)

© 2018 Larsen & Keller

⊟ Larsen & Keller

Published by Larsen and Keller Education,
5 Penn Plaza,
19th Floor,
New York, NY 10001, USA

Cataloging-in-Publication Data

Introductory biogeography / edited by Emma Darling.
   p. cm.
Includes bibliographical references and index.
ISBN 978-1-63549-696-3
1. Biogeography. 2. Biotic communities--Geographical distribution. 3. Biology.
I. Darling, Emma.
QH84 .I58 2018
578.09--dc23

For more information regarding Larsen and Keller Education and its products, please visit the publisher's website www.larsen-keller.com

Table of Contents

Preface

Earth inhabits many species of flora and fauna. The existence of these infinite and vastly different species depends on their varied geographical, natural, environmental and geological environment and space. The study of this distribution based on geological time and geographical space is called biogeography. It has two main branches namely, zoogeography and phytogeography. This book explores all the important aspects of biogeography in the present-day scenario. It is designed to provide deep insights about the subject to the readers. The topics included in the text are based on the most important areas of this vast subject. This textbook will serve as a reference to a broad spectrum of readers.

A short introduction to every chapter is written below to provide an overview of the content of the book:

Chapter 1 - The study of the distribution of organisms and their habitats is known as biogeography. The branch of biogeography which studies the distribution of plants can be termed as phytogeography. The chapter on biogeography offers an insightful focus, keeping in mind the complex subject matter; **Chapter 2** - Biogeographic realms are spatial regions where ecosystems share a similar biota. Afrotropical realm, Antarctic realm, Palearctic realm, Neotropical realm and Indomalayan realm are some of the recognized biogeographic realms. This chapter is an overview of the subject matter incorporating all the major aspects of biogeographic realm; **Chapter 3** - A biome is classified as an area where the animals and plants have common characteristics. The common characteristics are mainly influenced by the climatic conditions of the area. The topics discussed in the chapter are of great importance to broaden the existing knowledge on biome; **Chapter 4** - The fundamental concepts of biogeography are allopatric speciation, evolution, biological dispersal extinction and endemism. Allopatric speciation is the process that occurs when the population of the same species becomes secluded with each other resulting in genetic variations. This chapter discusses the fundamental concepts of biogeography in a critical manner providing key analysis to the subject matter; **Chapter 5** - Speciation is the process by which the population of the same species evolves into separate species. The types of speciation that occurs are peripatric speciation, sympatric speciation, ecological speciation, etc. This section has been carefully written to provide an easy understanding of the varied facets of speciation.

Finally, I would like to thank my fellow scholars who gave constructive feedback and my family members who supported me at every step.

Editor

An Introduction to Biogeography

The study of the distribution of organisms and their habitats is known as biogeography. The branch of biogeography which studies the distribution of plants can be termed as phytogeography. The chapter on biogeography offers an insightful focus, keeping in mind the complex subject matter.

Biogeography

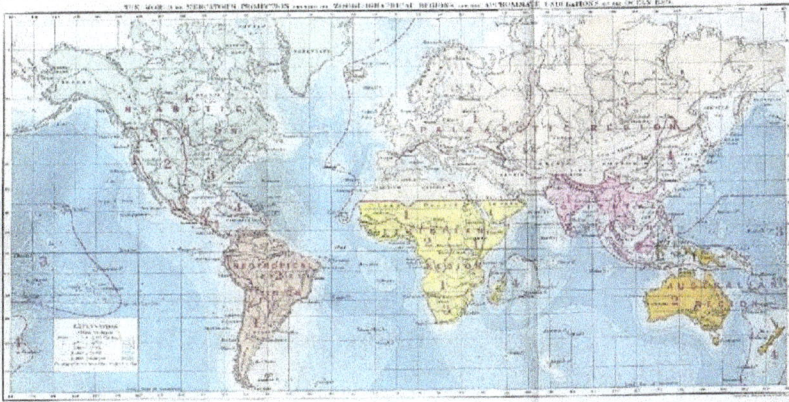

Frontispiece to Alfred Russel Wallace's book The Geographical Distribution of Animals

Biogeography is the study of the distribution of species and ecosystems in geographic space and through (geological) time. Organisms and biological communities often vary in a regular fashion along geographic gradients of latitude, elevation, isolation and habitat area. Phytogeography is the branch of biogeography that studies the distribution of plants. Zoogeography is the branch that studies distribution of animals.

Knowledge of spatial variation in the numbers and types of organisms is as vital to us today as it was to our early human ancestors, as we adapt to heterogeneous but geographically predictable environments. Biogeography is an integrative field of inquiry that unites concepts and information from ecology, evolutionary biology, geology, and physical geography.

Modern biogeographic research combines information and ideas from many fields, from the physiological and ecological constraints on organismal dispersal to geological and climatological phenomena operating at global spatial scales and evolutionary time frames.

The short-term interactions within a habitat and species of organisms describe the eco-logical application of biogeography. Historical biogeography describes the long-term, evolutionary periods of time for broader classifications of organisms. Early scientists, beginning with Carl Linnaeus, contributed to the development of biogeography as a sci-ence. Beginning in the mid-18th century, Europeans explored the world and discovered the biodiversity of life.

The scientific theory of biogeography grows out of the work of Alexander von Hum-boldt (1769–1859), Hewett Cottrell Watson (1804–1881), Alphonse de Candolle (1806–1893), Alfred Russel Wallace (1823–1913), Philip Lutley Sclater (1829–1913) and other biologists and explorers.

Introduction

The patterns of species distribution across geographical areas can usually be explained through a combination of historical factors such as: speciation, extinction, continental drift, and glaciation. Through observing the geographic distribution of species, we can see associated variations in sea level, river routes, habitat, and river capture. Addition-ally, this science considers the geographic constraints of landmass areas and isolation, as well as the available ecosystem energy supplies.

Over periods of ecological changes, biogeography includes the study of plant and ani-mal species in: their past and/or present living *refugium* habitat; their interim living sites; and/or their survival locales. As writer David Quammen put it, "...biogeography does more than ask *Which species?* and *Where*. It also asks *Why?* and, what is some-times more crucial, *Why not?*."

Modern biogeography often epmploys the use of Geographic Information Systems (GIS), to understand the factors affecting organism distribution, and to predict future trends in organism distribution. Often mathematical models and GIS are employed to solve ecological problems that have a spatial aspect to them.

Biogeography is most keenly observed on the world's islands. These habitats are often much more manageable areas of study because they are more condensed than larger ecosystems on the mainland. Islands are also ideal locations because they allow scien-tists to look at habitats that new invasive species have only recently colonized and can observe how they disperse throughout the island and change it. They can then apply their understanding to similar but more complex mainland habitats. Islands are very diverse in their biomes, ranging from the tropical to arctic climates. This diversity in habitat allows for a wide range of species study in different parts of the world.

One scientist who recognized the importance of these geographic locations was Charles Darwin, who remarked in his journal "The Zoology of Archipelagoes will be well worth examination". Two chapters in *On the Origin of Species* were devoted to geographical distribution.

History

18th Century

The first discoveries that contributed to the development of biogeography as a science began in the mid-18th century, as Europeans explored the world and discovered the biodiversity of life. During the 18th century most views on the world were shaped around religion and for many natural theologists, the bible. Carl Linnaeus, in the mid-18th century, initiated the ways to classify organisms through his exploration of undiscovered territories. When he noticed that species were not as perpetual as he believed, he developed the Mountain Explanation to explain the distribution of biodiversity. When Noah's ark landed on Mount Ararat and the waters receded, the animals dispersed throughout different elevations on the mountain. This showed different species in different climates proving species were not constant. Linnaeus' findings set a basis for ecological biogeography. Through his strong beliefs in Christianity, he was inspired to classify the living world, which then gave way to additional accounts of secular views on geographical distribution. He argued that the structure of an animal was very closely related to its physical surroundings. This was important to a George Louis Buffon's rival theory of distribution.

Edward O. Wilson, a prominent biologist and conservationist, coauthored *The Theory of Island Biogeography* and helped to start much of the research that has been done on this topic since the work of Watson and Wallace almost a century before

Closely after Linnaeus, Georges-Louis Leclerc, Comte de Buffon observed shifts in climate and how species spread across the globe as a result. He was the first to see different groups of organisms in different regions of the world. Buffon saw similarities between some regions which led him to believe that at one point continents were connected and then water separated them and caused differences in species. His hypotheses were described by his books, Histoire Naturelle, and Générale et Particulière, in which he argued that varying geographical regions would have different forms of life. This was inspired by his observations comparing the Old and New World, as he determined distinct variations of species from the two regions. Buffon believed there was a single species creation event, and that different regions of the world were homes for varying species, which is an alternate view than that of Linnaeus. Buffon's Law eventually became a principle of biogeography by explaining how similar environments were habitats for comparable types of organisms. Buffon also studied fossils which led him to believe that the earth was over tens of thousands of years old, and that humans had not lived there long in comparison to the age of the earth.

Following this period of exploration came the Age of Enlightenment in Europe, which attempted to explain the patterns of biodiversity observed by Buffon and Linnaeus. At the end of the 18th century, Alexander von Humboldt, known as the "founder of plant geography", developed the concept of physique generale to demonstrate the unity of science and how species fit together. As one of the first to contribute empirical data to the science of biogeography through his travel as an explorer, he observed differences in climate and vegetation. The earth was divided into regions which he defined as tropical, temperate, and arctic and within these regions there were similar forms of vegetation. This ultimately enabled him to create the isotherm, which allowed scientists to see patterns of life within different climates. He contributed his observations to findings of botanical geography by previous scientists, and sketched this description of both the biotic and abiotic features of the earth in his book, Cosmos.

Augustin de Candolle contributed to the field of biogeography as he observed species competition and the several differences that influenced the discovery of the diversity of life. He was a Swiss botanist and created the first Laws of Botanical Nomenclature in his work, Prodromus. He discussed plant distribution and his theories eventually had a great impact on Charles Darwin, who was inspired to consider species adaptations and evolution after learning about botanical geography. De Candolle was the first to describe the differences between the small-scale and large-scale distribution patterns of organisms around the globe.

19th Century

In the 19th century, several additional scientists contributed new theories to further develop the concept of biogeography. Charles Lyell, being one of the first contributors in the 19th century, developed the Theory of Uniformitarianism after studying fossils. This theory explained how the world was not created by one sole catastrophic event, but instead from numerous creation events and locations. Uniformitarianism also introduced the idea that the Earth was actually significantly older than was previously accepted. Using this knowledge, Lyell concluded that it was possible for species to go extinct. Since he noted that earth's climate changes, he realized that species distribution must also change accordingly. Lyell argued that climate changes complemented vegetation changes, thus connecting the environmental surroundings to varying species. This largely influenced Charles Darwin in his development of the theory of evolution.

Charles Darwin was a natural theologist who studied around the world, and most importantly in the Galapagos Islands. Darwin introduced the idea of natural selection, as he theorized against previously accepted ideas that species were static or unchanging. His contributions to biogeography and the theory of evolution were different from those of other explorers of his time, because he developed a mechanism to describe the ways that species changed. His influential ideas include the development of theories regarding the struggle for existence and natural selection. Darwin's theories started a biological segment to biogeography and empirical studies, which enabled future scientists to develop ideas about the geographical distribution of organisms around the globe.

Alfred Russel Wallace studied the distribution of flora and fauna in the Amazon Basin and the Malay Archipelago in the mid-19th century. His research was essential to the further development of biogeography, and he was later nicknamed the "father of Biogeography". Wallace conducted fieldwork researching the habits, breeding and migration tendencies, and feeding behavior of thousands of species. He studied butterfly and bird distributions in comparison to the presence or absence of geographical barriers. His observations led him to conclude that the number of organisms present in a community was dependent on the amount of food resources in the particular habitat. Wallace believed species were dynamic by responding to biotic and abiotic factors. He and Philip Sclater saw biogeography as a source of support for the theory of evolution as they used Darwin's conclusion to explain how biogeography was similar to a record of species inheritance. Key findings, such as the sharp difference in fauna either side of the Wallace Line, and the sharp difference that existed between North and South America prior to their relatively recent faunal interchange, can only be understood in this light. Otherwise, the field of biogeography would be seen as a purely descriptive one.

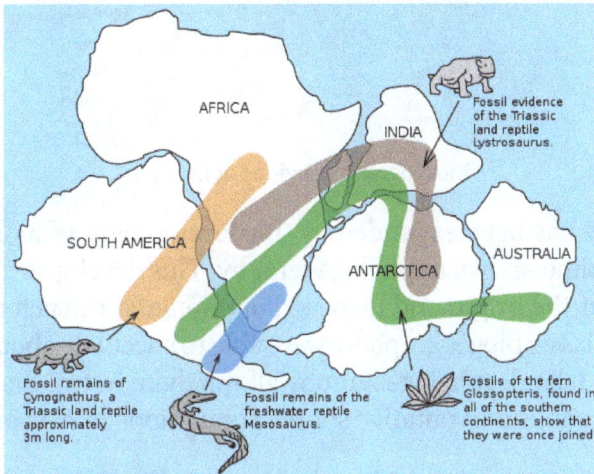

Schematic distribution of fossils on Pangea according to Wegener

20th and 21st Century

Moving on to the 20th century, Alfred Wegener introduced the Theory of Continental Drift in 1912, though it was not widely accepted until the 1960s. This theory was revolutionary because it changed the way that everyone thought about species and their distribution around the globe. The theory explained how continents were formerly joined together in one large landmass, Pangea, and slowly drifted apart due to the movement of the plates below Earth's surface. The evidence for this theory is in the geological similarities between varying locations around the globe, fossil comparisons from different continents, and the jigsaw puzzle shape of the landmasses on Earth. Though Wegener did not know the mechanism of this concept of Continental Drift, this contribution to the study of biogeography was significant in the way that it shed light on the importance of environmental and geographic similarities or differences as a result of climate and other pressures on the planet.

The publication of *The Theory of Island Biogeography* by Robert MacArthur and E.O. Wilson in 1967 showed that the species richness of an area could be predicted in terms of such factors as habitat area, immigration rate and extinction rate. This added to the long-standing interest in island biogeography. The application of island biogeography theory to habitat fragments spurred the development of the fields of conservation biology and landscape ecology.

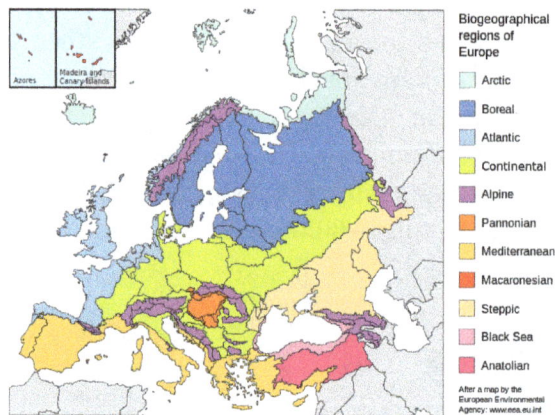

Biogeographic regions of Europe

Classic biogeography has been expanded by the development of molecular systematics, creating a new discipline known as phylogeography. This development allowed scientists to test theories about the origin and dispersal of populations, such as island endemics. For example, while classic biogeographers were able to speculate about the origins of species in the Hawaiian Islands, phylogeography allows them to test theories of relatedness between these populations and putative source populations in Asia and North America.

Biogeography continues as a point of study for many life sciences and geography students worldwide, however it may be under different broader titles within institutions such as ecology or evolutionary biology.

In recent years, one of the most important and consequential developments in biogeography has been to show how multiple organisms, including mammals like monkeys and reptiles like lizards, overcame barriers such as large oceans that many biogeographers formerly believed were impossible to cross.

Modern Applications of Biogeography

Biogeography now incorporates many different fields including but not limited to physical geography, geology, botany and plant biology, zoology, and general biology. A biogeographer's main focus is on what environmental factors and what the influence of humans do to the distribution of the specific species of study. In terms of applications of biogeography as a science today, technological advances have allowed satellite imaging and processing of the Earth. Two main types of satellite imaging that are important within modern biogeography are Global Production Efficiency Model (GLO-PEM) and General Information Sensing (GIS). GLO-PEM uses satellite-imaging gives "repetitive, spatially contiguous, and time specific observations of vegetation." These observations are on a global scale. GIS can show certain processes on the earth's surface like whale locations, sea surface temperatures, and bathymetry. Current scientists also use coral reefs to delve into the history of biogeography through the fossilized reefs.

Paleobiogeography

Paleobiogeography goes one step further to include paleogeographic data and considerations of plate tectonics. Using molecular analyses and corroborated by fossils, it has been possible to demonstrate that perching birds evolved first in the region of Australia or the adjacent Antarctic (which at that time lay somewhat further north and had a temperate climate). From there, they spread to the other Gondwanan continents and Southeast Asia – the part of Laurasia then closest to their origin of dispersal – in the late Paleogene, before achieving a global distribution in the early Neogene. Not knowing the fact that at the time of dispersal, the Indian Ocean was much narrower than it is today, and that South America was closer to the Antarctic, one would be hard pressed to explain the presence of many "ancient" lineages of perching birds in Africa, as well as the mainly South American distribution of the suboscines.

Paleobiogeography also helps constrain hypotheses on the timing of biogeographic events such as vicariance and geodispersal, and provides unique information on the formation of regional biotas. For example, data from species-level phylogenetic and biogeographic studies tell us that the Amazonian fish fauna accumulated in increments over a period of tens of millions of years, principally by means of allopatric speciation, and in an arena extending over most of the area of tropical South America (Albert & Reis 2011). In other words, unlike some of the well-known insular faunas (Galapagos finches, Hawaiian drosophilid flies, African rift lake cichlids), the species-rich Amazonian ichthyofauna is not the result of recent adaptive radiations.

For freshwater organisms, landscapes are divided naturally into discrete drainage basins by watersheds, episodically isolated and reunited by erosional processes. In regions like the Amazon Basin with an exceptionally low (flat) topographic relief, the many waterways have had a highly reticulated history over geological time. In such a context, stream capture is an important factor affecting the evolution and distribution of freshwater organisms. Stream capture occurs when an upstream portion of one river drainage is diverted to the downstream portion of an adjacent basin. This can happen as a result of tectonic uplift (or subsidence), natural damming created by a landslide, or headward or lateral erosion of the watershed between adjacent basins.

Concepts and Fields

Biogeography is a synthetic science, related to geography, biology, soil science, geology, climatology, ecology and evolution.

Some fundamental concepts in biogeography include:

- allopatric speciation – the splitting of a species by evolution of geographically isolated populations

- evolution – change in genetic composition of a population

- extinction – disappearance of a species

- dispersal – movement of populations away from their point of origin, related to migration

- endemic areas

- geodispersal – the erosion of barriers to biotic dispersal and gene flow, that permit range expansion and the merging of previously isolated biotas

- range and distribution

- vicariance – the formation of barriers to biotic dispersal and gene flow, that tend to subdivide species and biotas, leading to speciation and extinction

Comparative Biogeography

The study of comparative biogeography can follow two main lines of investigation:

- Systematic biogeography is the study of biotic area relationships, their distribution, and hierarchical classification.

- Evolutionary biogeography is the proposal of evolutionary mechanisms responsible for organismal distributions. Possible mechanisms include widespread taxa disrupted by continental break-up or individual episodes of long-distance movement.

Biogeographic Regionalisations

There are many types of biogeographic units used in biogeographic regionalisation schemes, as there are many criteria (species composition, physiognomy, ecological aspects) and hierarchization schemes: biogeographic realms (or ecozones), bioregions (*sensu stricto*), ecoregions, zoogeographical regions, floristic regions, vegetation types, biomes, etc.

The terms "biogeographic unit" (Calow, 1998), "biogeographic area" (Ebach et al., 2008) or "bioregion" (*sensu lato*, Vilhena & Antonelli, 2015) can be used for these categories, regardless of rank.

Recently, a International Code of Area Nomenclature was proposed for biogeography.

Phytogeography

Phytogeography or botanical geography is the branch of biogeography that is concerned with the geographic distribution of plant species and their influence on the earth's surface. Phytogeography is concerned with all aspects of plant distribution, from the controls on the distribution of individual species ranges (at both large and small scales) to the factors that govern the composition of entire communities and floras. Geobotany, by contrast, focuses on the geographic space's influence on plants.

Fields

Phytogeography is part of a more general science known as biogeography. Phytogeographers are concerned with patterns and process in plant distribution. Most of the major questions and kinds of approaches taken to answer such questions are held in common between phyto- and zoogeographers.

Phytogeography in wider sense (or geobotany, in German literature) encompasses four fields, according with the focused aspect, environment, flora (taxa), vegetation (plant community) and origin, respectively:

- plant ecology (or mesology - however, the physiognomic-ecological approach on vegetation and biome study are also generally associated with this field).

- plant geography (or phytogeography in strict sense, chorology, floristics).

- plant sociology (or phytosociology, synecology - however, this field doesn't prescind from flora study, as its approach to study vegetation relies upon a fundamental unit, the plant association, which is defined upon flora).

- historical plant geography (or paleobotany, paleogeobotany)

Phytogeography is often divided into two main branches: ecological phytogeography and historical phytogeography. The former investigates the role of current day biotic and abiotic interactions in influencing plant distributions; the latter are concerned with historical reconstruction of the origin, dispersal, and extinction of taxa.

Overview

The basic data element of phytogeography are specimen records. These are collected individual plants like this one, a Cinnamon Fern, collected in the Smokey Mountains of North Carolina.

The basic data elements of phytogeography are occurrence records (presence or absence of a species) with operational geographic units such as political units or geographical coordinates. These data are often used to construct phytogeographic provinces (floristic provinces) and elements.

The questions and approaches in phytogeography are largely shared with zoogeography, except zoogeography is concerned with animal distribution rather than plant distribution. The term phytogeography itself suggests a broad meaning. How the term is actually applied by practicing scientists is apparent in the way periodicals use the term. The *American Journal of Botany*, a monthly primary research journal, frequently publishes a section titled "Systematics, Phytogeography, and Evolution." Topics covered in the *American Journal of Botany*'s "Systematics and Phytogeography" section include phylogeography, distribution of genetic variation and, historical biogeography, and general plant species distribution patterns. Biodiversity patterns are not heavily covered.

History

Phytogeography has a long history. One of the subjects earliest proponents was Prussian naturalist Alexander von Humboldt, who is often referred to as the "father of phytogeography". Von Humboldt advocated a quantitative approach to phytogeography that has characterized modern plant geography.

Gross patterns of the distribution of plants became apparent early on in the study of plant geography. For example, Alfred Russel Wallace, co-discoverer of the principle of natural selection, discussed the Latitudinal gradients in species diversity, a pattern observed in other organisms as well. Much research effort in plant geography has since then been devoted to understanding this pattern and describing it in more detail.

An 1814 self-portrait in Paris of Alexander von Humboldt. Humboldt is often referred to as the "father of phytogeography".

In 1890, the United States Congress passed an act that appropriated funds to send expeditions to discover the geographic distributions of plants (and animals) in the United States. The first of these was The Death Valley Expedition, including Frederick Vernon Coville, Frederick Funston, Clinton Hart Merriam, and others.

Research in plant geography has also been directed to understanding the patterns of adaptation of species to the environment. This is done chiefly by describing geographical patterns of trait/environment relationships. These patterns termed ecogeographical rules when applied to plants represent another area of phytogeography. Recently, a new field termed macroecology has developed, which focuses on broad-scale (in both time and space) patterns and phenomena in ecology. Macroecology focuses as much on other organisms as plants.

Floristic Regions

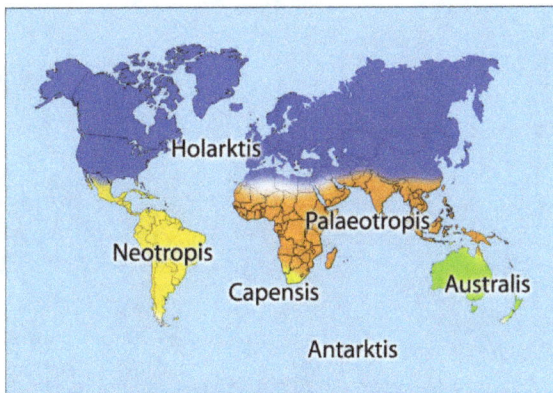

Good (1947) floristic kingdoms

Floristics is a study of the flora of some territory or area. Traditional phytogeography concerns itself largely with floristics and floristic classification.

References

- Ebach, M.C., Morrone, J.J. Parenti, L.R. & Viloria Á.L. (2008). International Code of Area Nomenclature. Journal of Biogeography 35 (7): 1153–1157,[5]

- "Remote Sensing Data and Information." Remote Sensing Data and Information. "Archived copy". Archived from the original on 2014-04-27. Retrieved 2014-04-28. (accessed April 28, 2014)

- Stephen D. Prince and Samuel N. Goward. "Global Primary Production: A Remote Sensing Approach" Journal of Biogeography, Vol. 22, No. 4/5, Terrestrial Ecosystem Interactions with Global Change, Volume 2 (Jul. - Sep., 1995), pp. 815-835

- Nicolson, D.H. (1991). "A History of Botanical Nomenclature". Annals of the Missouri Botanical Garden. 78 (1): 33–56. doi:10.2307/2399589

- Jønsson, Knud A. & Fjeldså, Jon (2006). Determining biogeographical patterns of dispersal and diversification in oscine passerine birds in Australia, Southeast Asia and Africa. Journal of Biogeography 33(7): 1155–1165. doi:10.1111/j.1365-2699.2006.01507

2

Biogeographic Realm: An Overview

Biogeographic realms are spatial regions where ecosystems share a similar biota. Afrotropical realm, Antarctic realm, Palearctic realm, Neotropical realm and Indomalayan realm are some of the recognized biogeographic realms. This chapter is an overview of the subject matter incorporating all the major aspects of biogeographic realm.

Biogeographic Realm

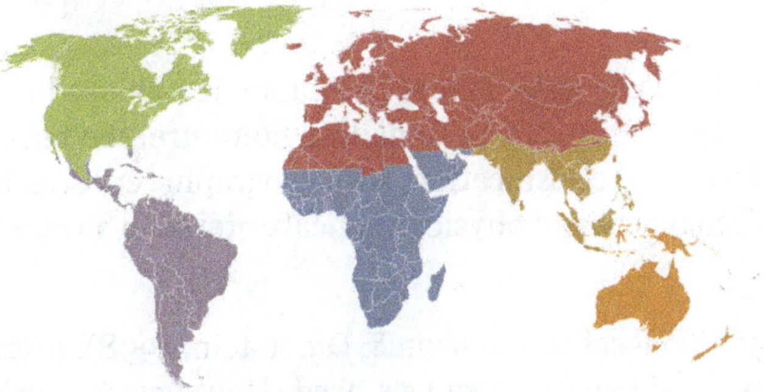

6 of the 8 biogeographic realms according to the WWF

Nearctic
Palearctic
Afrotropic
Indomalaya
Australasia
Neotropic
Oceania and Antarctic ecozones not shown.

A biogeographic realm or ecozone is the broadest biogeographic division of the Earth's land surface, based on distributional patterns of terrestrial organisms. They are subdivided in ecoregions, which are classified in biomes or habitat types.

The realms delineate large areas of the Earth's surface within which organisms have been evolving in relative isolation over long periods of time, separated from one another by geographic features, such as oceans, broad deserts, or high mountain ranges, that constitute barriers to migration. As such, biogeographic realms designations are used to indicate general groupings of organisms based on their shared biogeography. Biogeographic realms correspond to the floristic kingdoms of botany or zoogeographic regions of zoology.

Biogeographic realms are characterized by the evolutionary history of the organisms they contain. They are distinct from biomes, also known as major habitat types, which are divisions of the Earth's surface based on *life form*, or the adaptation of animals, fungi, micro-organisms and plants to climatic, soil, and other conditions. Biomes are characterized by similar climax vegetation. Each realm may include a number of different biomes. A tropical moist broadleaf forest in Central America, for example, may be similar to one in New Guinea in its vegetation type and structure, climate, soils, etc., but these forests are inhabited by animals, fungi, micro-organisms and plants with very different evolutionary histories.

The patterns of distribution of living organisms in the world's biogeographic realms were shaped by the process of plate tectonics, which has redistributed the world's land masses over geological history.

Concept History

The "biogeographic realms" of Udvardy (1975) were defined based on taxonomic composition. The rank corresponds more or less to the floristic kingdoms and zoogeographic regions.

The usage of the term "ecozone" is more variable. It was used originally in stratigraphy (Vella, 1962, Hedberg, 1971). In Canadian literature, the term was used by Wiken (1986) in macro level land classification, with geographic criteria. Later, Schültz (1988) would use it with ecological and physiognomical criteria, in a way similar to the concept of biome.

In the Global 200/WWF scheme (Olson & Dinerstein, 1998), originally the term "biogeographic realm" in Udvardy sense was used. However, in a scheme of BBC, it was replaced by the term "ecozone".

Terrestrial Biogeographic Realms

Udvardy (1975) biogeographic realms

The hierarchy of the scheme is (with early replaced terms in parenthesis):

- biogeographic realm (= biogeographic regions and subregions), with 8 categories

  o biogeographic province (= biotic province), with 193 categories, each characterized by a major biome or biome-complex

    □ biome, with 14 types

The realms and provinces of the scheme are:

(1) Nearctic realm

(2) Palaearctic realm

(3) Africotropical realm

(4) Indomalayan realm

(5) Oceanian realm

(6) Australian realm

(7) Antarctic realm

(8) Neotropical realm

WWF / Global 200 biogeographic realms (BBC "ecozones")

The World Wildlife Fund scheme (Olson & Dinerstein, 1998, Olson et al., 2001) is broadly similar to Miklos Udvardy's system, the chief difference being the delineation of the Australasian realm relative to the Antarctic, Oceanic, and Indomalayan realms. In the WWF system, The Australasia realm includes Australia, Tasmania, the islands of Wallacea, New Guinea, the East Melanesian islands, New Caledonia, and New Zealand. Udvardy's Australian realm includes only Australia and Tasmania; he places Wallacea in the Indomalayan Realm, New Guinea, New Caledonia, and East Melanesia in the Oceanian Realm, and New Zealand in the Antarctic Realm.

Biogeographic realm	Area		Notes
	million square kilometres	million square miles	
Palearctic	54.1	20.9	including the bulk of Eurasia and North Africa
Nearctic	22.9	8.8	including most of North America
Afrotropic	22.1	8.5	including Trans-Saharan Africa and Arabia
Neotropic	19.0	7.3	including South America, Central America, and the Caribbean
Australasia	7.6	2.9	including Australia, New Guinea, New Zealand, and neighbouring islands. The northern boundary of this zone is known as the Wallace line.
Indo-Malaya	7.5	2.9	including the Indian subcontinent, Southeast Asia, and southern China
Oceania	1.0	0.39	including Polynesia (except New Zealand), Micronesia, and the Fijian Islands
Antarctic	0.3	0.12	including Antarctica.

The Palearctic and Nearctic are sometimes grouped into the Holarctic realm.

Morrone (2015) biogeographic kingdoms

Following the nomenclatural conventions set out in the International Code of Area Nomenclature, Morrone (2015) defined the next biogeographic kingdoms (or realms) and regions:

- Holarctic kingdom Heilprin (1887)

    o   Nearctic region Sclater (1858)

    o   Palearctic region Sclater (1858)

- Holotropical kingdom Rapoport (1968)

    o   Neotropical region Sclater (1858)

    o   Ethiopian region Sclater (1858)

    o   Oriental region Wallace (1876)

- Austral kingdom Engler (1899)

    o   Cape region Grisebach (1872)

    o   Andean region Engler (1882)

    o   Australian region Sclater (1858)

    o   Antarctic region Grisebach (1872)

- Transition zones:

    o   Mexican transition zone (Nearctic–Neotropical transition)

    o   Saharo-Arabian transition zone (Palearctic–Ethiopian transition)

    o   Chinese transition zone (Palearctic–Oriental transition zone transition)

    o   Indo-Malayan, Indonesian or Wallace's transition zone (Oriental–Australian transition)

    o   South American transition zone (Neotropical–Andean transition)

Freshwater Biogeographic Realms

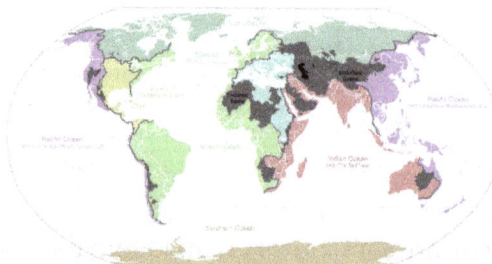

The applicability of Udvardy (1975) scheme to most freshwater taxa is unresolved.

The drainage basins of the principal oceans and seas of the world are marked by continental divides. The grey areas are endorheic basins that do not drain to the ocean.

Marine Biogeographic Realms

According to Briggs (1995):

- Indo-West Pacific region
- Eastern Pacific region
- Western Atlantic region
- Eastern Atlantic region
- Southern Australian region
- Northern New Zealand region
- Western South America region
- Eastern South America region
- Southern Africa region
- Mediterranean–Atlantic region
- Japan region
- Carolina region
- California region
- Tasmanian region
- Southern New Zealand region
- Antipodean region
- Subantarctic region
- Magellan region
- Eastern Pacific Boreal region
- Western Atlantic Boreal region
- Eastern Atlantic Boreal region
- Antarctic region
- Arctic region

According to the WWF scheme (Spalding, 2007):

- Arctic realm
- Temperate Northern Atlantic realm
- Temperate Northern Pacific realm
- Tropical Atlantic realm
- Western Indo-Pacific realm
- Central Indo-Pacific realm
- Eastern Indo-Pacific realm
- Tropical Eastern Pacific realm

- Temperate South America realm

- Temperate Southern Africa realm

- Temperate Australasia realm

- Southern Ocean realm

Ecoregion

A map location of the Amazon rainforest ecoregions. The yellow line encloses the ecoregions per the World Wide Fund for Nature.

An ecoregion (ecological region) is an ecologically and geographically defined area that is smaller than a bioregion, which in turn is smaller than an ecozone. All three of these are either less or greater than an ecosystem. Ecoregions cover relatively large areas of land or water, and contain characteristic, geographically distinct assemblages of natural communities and species. The biodiversity of flora, fauna and ecosystems that characterise an ecoregion tends to be distinct from that of other ecoregions. In theory, biodiversity or conservation ecoregions are relatively large areas of land or water where the probability of encountering different species and communities at any given point remains relatively constant, within an acceptable range of variation (largely undefined at this point).

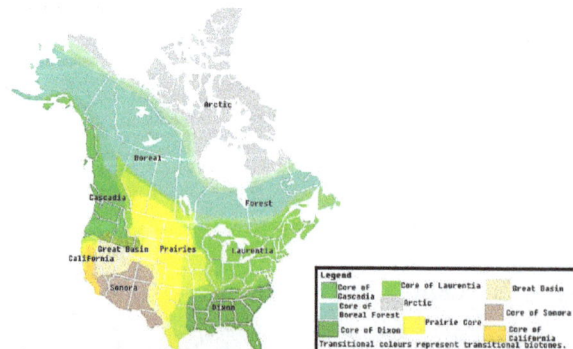

A map of North America's bioregions

Three caveats are appropriate for all bio-geographic mapping approaches. Firstly, no single bio-geographic framework is optimal for all taxa. Ecoregions reflect the best compromise for as many taxa as possible. Secondly, ecoregion boundaries rarely form abrupt edges; rather, ecotones and mosaic habitats bound them. Thirdly, most ecoregions contain habitats that differ from their assigned biome. Biogeographic provinces may originate due to various barriers. Some physical (plate tectonics, topographic highs), some climatic (latitudinal variation, seasonal range) and some ocean chemical related (salinity, oxygen levels).

History

The history of the term is somewhat vague as it was used in many contexts: forest classifications (Loucks, 1962), biome classifications (Bailey, 1976, 2014), biogeographic classifications (WWF/Global 200 scheme of Olson & Dinerstein, 1998), etc.

The concept of ecoregion of Bailey gives more importance to ecological criteria, while the WWF concept gives more importance to biogeography, that is, distribution of distinct biotas.

Definition and categorization

The Ötztal Alps, a mountain range in the central Alps of Europe, are part of the Central Eastern Alps, and can both be termed as ecoregions.

A conifer forest in the Swiss Alps (National Park).

An ecoregion is a "recurring pattern of ecosystems associated with characteristic combinations of soil and landform that characterise that region" Omernik (2004), elaborates on this by defining ecoregions as: "areas within which there is spatial coincidence in characteristics of geographical phenomena associated with differences in the quality, health, and integrity of ecosystems" "Characteristics of geographical phenomena" may include geology, physiography, vegetation, climate, hydrology, terrestrial and aquatic fauna, and soils, and may or may not include the impacts of human activity (e.g. land use patterns, vegetation changes). There is significant, but not absolute, spatial correlation among these characteristics, making the delineation of ecoregions an imperfect science. Another complication is that environmental conditions across an ecoregion boundary may change very gradually, e.g. the prairie-forest transition in the midwestern United States, making it difficult to identify an exact dividing boundary. Such transition zones are called ecotones.

Ecoregions can be categorized using an algorithmic approach or a holistic, "weight-of-evidence" approach where the importance of various factors may vary. An example of the algorithmic approach is Robert Bailey's work for the U.S. Forest Service, which uses a hierarchical classification that first divides land areas into very large regions based on climatic factors, and subdivides these regions, based first on dominant potential vegetation, and then by geomorphology and soil characteristics. The weight-of-evidence approach is exemplified by James Omernik's work for the United States Environmental Protection Agency, subsequently adopted (with modification) for North America by the Commission for Environmental Cooperation.

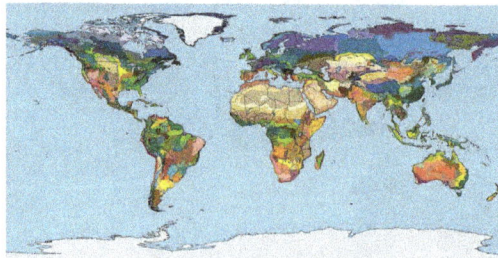

Terrestrial Ecoregions of the World (Olson et al. 2001, BioScience)

The intended purpose of ecoregion delineation may affect the method used. For example, the WWF ecoregions were developed to aid in biodiversity conservation planning, and place a greater emphasis than the Omernik or Bailey systems on floral and faunal differences between regions. The WWF classification defines an ecoregion as:

A large area of land or water that contains a geographically distinct assemblage of natural communities that:

(a) Share a large majority of their species and ecological dynamics;

(b) Share similar environmental conditions, and;

(c) Interact ecologically in ways that are critical for their long-term persistence.

According to WWF, the boundaries of an ecoregion approximate the original extent of the natural communities prior to any major recent disruptions or changes. WWF has identified 867 terrestrial ecoregions, and approximately 450 freshwater ecoregions across the Earth.

Importance

The use of the term ecoregion is an outgrowth of a surge of interest in ecosystems and their functioning. In particular, there is awareness of issues relating to spatial scale in the study and management of landscapes. It is widely recognized that interlinked ecosystems combine to form a whole that is "greater than the sum of its parts". There are many attempts to respond to ecosystems in an integrated way to achieve "multi-functional" landscapes, and various interest groups from agricultural researchers to conservationists are using the "ecoregion" as a unit of analysis.

The "Global 200" is the list of ecoregions identified by WWF as priorities for conservation.

Ecologically based movements like bioregionalism maintain that ecoregions, rather than arbitrarily defined political boundaries, provide a better foundation for the formation and governance of human communities, and have proposed ecoregions and watersheds as the basis for bioregional democracy initiatives.

Terrestrial

WWF terrestrial ecoregions

Terrestrial ecoregions are land ecoregions, as distinct from freshwater and marine ecoregions. In this context, *terrestrial* is used to mean "of land" (soil and rock), rather than the more general sense "of Earth" (which includes land and oceans).

WWF (World Wildlife Fund) ecologists currently divide the land surface of the Earth into 8 major ecozones containing 867 smaller terrestrial ecoregions. The WWF effort is a synthesis of many previous efforts to define and classify ecoregions. Many consider this classification to be quite decisive, and some propose these as stable borders for bioregional democracy initiatives.

The eight terrestrial ecozones follow the major floral and faunal boundaries, identified by botanists and zoologists, that separate the world's major plant and animal communities. Ecozone boundaries generally follow continental boundaries, or major barriers to plant and animal distribution, like the Himalayas and the Sahara. The boundaries of ecoregions are often not as decisive or well recognized, and are subject to greater disagreement.

Ecoregions are classified by biome type, which are the major global plant communities determined by rainfall and climate. Forests, grasslands (including savanna and shrubland), and deserts (including xeric shrublands) are distinguished by climate (tropical and subtropical vs. temperate and boreal climates) and, for forests, by whether the trees are predominantly conifers (gymnosperms), or whether they are predominantly broadleaf (Angiosperms) and mixed (broadleaf and conifer). Biome types like Mediterranean forests, woodlands, and scrub; tundra; and mangroves host very distinct ecological communities, and are recognized as distinct biome types as well.

Marine

View of Earth, taken in 1972 by the Apollo 17 crew. Approximately 72% of the Earth's surface (an area of some 361 million square kilometers) consists of ocean.

Marine ecoregions are: "Areas of relatively homogeneous species composition, clearly distinct from adjacent systems....In ecological terms, these are strongly cohesive units, sufficiently large to encompass ecological or life history processes for most sedentary species." They have been defined by The Nature Conservancy (TNC) and World Wildlife Fund (WWF) to aid in conservation activities for marine ecosystems. Forty-three priority marine ecoregions were delineated as part of WWF's Global 200 efforts. The scheme used to designate and classify marine ecoregions is analogous to that used for terrestrial ecoregions. Major habitat types are identified: polar, temperate shelves and seas, temperate upwelling, tropical upwelling, tropical coral, pelagic (trades and westerlies), abyssal, and hadal (ocean trench). These correspond to the terrestrial biomes.

The Global 200 classification of marine ecoregions is not developed to the same level of detail and comprehensiveness as that of the terrestrial ecoregions; only the priority conservation areas are listed.

In 2007, TNC and WWF refined and expanded this scheme to provide a system of comprehensive near shore (to 200 meters depth) Marine Ecoregions of the World (MEOW). The 232 individual marine ecoregions are grouped into 62 marine provinces, which in turn group into 12 marine realms, which represent the broad latitudinal divisions of polar, temperate, and tropical seas, with subdivisions based on ocean basins (except for the southern hemisphere temperate oceans, which are based on continents).

Major biogeographic realms, analogous to the eight terrestrial ecozones, represent large regions of the ocean basins: Arctic, Temperate Northern Atlantic, Temperate Northern Pacific, Tropical Atlantic, Western Indo-Pacific, Central Indo-Pacific, Eastern Indo-Pacific, Tropical Eastern Pacific, Temperate South America, Temperate Southern Africa, Temperate Australasia, Southern Ocean.

A similar system of identifying areas of the oceans for conservation purposes is the system of large marine ecosystems (LMEs), developed by the US National Oceanic and Atmospheric Administration (NOAA).

Freshwater

The Amazon River in Brazil.

A freshwater ecoregion is a large area encompassing one or more freshwater systems that contains a distinct assemblage of natural freshwater communities and species. The freshwater species, dynamics, and environmental conditions within a given ecoregion are more similar to each other than to those of surrounding ecoregions and together form a conservation unit. Freshwater systems include rivers, streams, lakes, and wetlands. Freshwater ecoregions are distinct from terrestrial ecoregions, which identify biotic communities of the land, and marine ecoregions, which are biotic communities of the oceans.

A new map of Freshwater Ecoregions of the World, released in 2008, has 426 ecoregions covering virtually the entire non-marine surface of the earth.

World Wildlife Fund (WWF) identifies twelve major habitat types of freshwater ecoregions: Large lakes, large river deltas, polar freshwaters, montane freshwaters, temperate coastal rivers, temperate floodplain rivers and wetlands, temperate upland rivers, tropical and subtropical coastal rivers, tropical and subtropical floodplain rivers and wetlands, tropical and subtropical upland rivers, xeric freshwaters and endorheic basins, and oceanic islands. The freshwater major habitat types reflect groupings of ecoregions with similar biological, chemical, and physical characteristics and are roughly equivalent to biomes for terrestrial systems.

The Global 200, a set of ecoregions identified by WWF whose conservation would achieve the goal of saving a broad diversity of the Earth's ecosystems, includes a number of areas highlighted for their freshwater biodiversity values. The Global 200 preceded Freshwater Ecoregions of the World and incorporated information from regional freshwater ecoregional assessments that had been completed at that time.

Afrotropical Realm

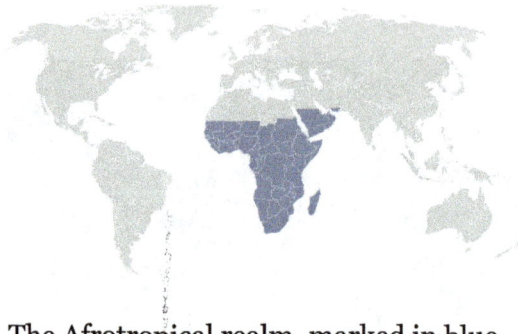

The Afrotropical realm, marked in blue

The Afrotropical realm is one of the Earth's eight biogeographic realms. It includes Africa south of the Sahara Desert, the southern and eastern fringes of the Arabian Peninsula, the island of Madagascar, southern Iran and extreme southwestern Pakistan, and the islands of the western Indian Ocean. It was formerly known as the Ethiopian Zone or Ethiopian Region.

Major Ecological Regions

Most of the Afrotropic, with the exception of Africa's southern tip, has a tropical climate. A broad belt of deserts, including the Atlantic and Sahara deserts of northern Africa and the Arabian Desert of the Arabian Peninsula, separate the Afrotropic from the Palearctic realm, which includes northern Africa and temperate Eurasia.

Sahel and Sudan

South of the Sahara, two belts of tropical grassland and savanna run east and west across the continent, from the Atlantic Ocean to the Ethiopian Highlands. Immediately south of the Sahara lies the Sahel belt, a transitional zone of semi-arid short grassland and Acacia savanna. Rainfall increases further south in the Sudanian Savanna, also known simply as the Sudan, a belt of taller grasslands and savannas. The Sudanian Savanna is home to two great flooded grasslands, the Sudd wetland in South Sudan, and the Niger Inland Delta in Mali. The forest-savanna mosaic is a transitional zone between the grasslands and the belt of tropical moist broadleaf forests near the equator.

Southern Arabian Woodlands

South Arabia, expressed as being mostly Yemen and parts of western Oman and southwestern Saudi Arabia, has few permanent forests. Some of the notable are Jabal Bura', Jabal Raymah, and Jabal Badaj in the Yemeni highland escarpment, and the seasonal forests in eastern Yemen and the Dhofar region of Oman. Other woodlands scatter the land and are very small and are predominantly juniper or acacia forests.

Forest Zone

The forest zone, a belt of lowland tropical moist broadleaf forests, runs across most of equatorial Africa's intertropical convergence zone. The Upper Guinean forests of West Africa extend along the coast from Guinea to Togo. The Dahomey Gap, a zone of forest-savanna mosaic that reaches to the coast, separates the Upper Guinean forests from the Lower Guinean forests, which extend along the Gulf of Guinea from eastern Benin through Cameroon and Gabon to the western Democratic Republic of the Congo. The largest tropical forest zone in Africa is the Congolian forests of the Congo Basin in Central Africa. A belt of tropical moist broadleaf forest also runs along the Indian Ocean coast, from southern Somalia to South Africa.

East African Grasslands and Savannas

- Acacia-Commiphora grasslands

- Serengeti

Eastern Africa's Highlands

Afromontane region, from the Ethiopian Highlands to the Drakensberg Mountains of South Africa, including the East African Rift. Distinctive flora, including *Podocarpus* and *Afrocarpus*, as well as giant *Lobelias* and *Senecios*.

- Ethiopian Highlands

- Albertine rift montane forests

- East African montane forests and Eastern Arc forests

Southern African woodlands, savannas, and grasslands

Southern Africa as described in Plant Taxonomic Database Standards No. 2

- Miombo woodlands
- Zambezian Mopane and Baikiaea woodlands
- Bushveld

Deserts of Southern Africa

- Namib Desert
- Kalahari Desert
- Karoo
- Tankwa Karoo
- Richtersveld

Cape Floristic Region

The Cape floristic region, at Africa's southern tip, is a Mediterranean climate region that is home to a significant number of endemic taxa, as well as to plant families like the proteas (*Proteaceae*) that are also found in the Australasian realm.

Madagascar and the Indian Ocean Islands

Madagascar and neighboring islands form a distinctive sub-region of the realm, with numerous endemic taxa like the lemurs. Madagascar and the Seychelles are old pieces of the ancient supercontinent of Gondwana, and broke away from Africa millions of years ago. Other Indian Ocean islands, like the Comoros and Mascarene Islands, are volcanic islands that formed more recently. Madagascar contains several important biospheres, as its biodiversity and ratio of endemicism is extremely high.

Endemic plants and animals

Plants

The Afrotropical realm is home to a number of endemic plant families. Madagascar and the Indian Ocean Islands are home to ten endemic families of flowering plants; eight are endemic to Madagascar (Asteropeiaceae, Didymelaceae, Didiereaceae, Kaliphoraceae, Melanophyllaceae, Physenaceae, Sarcolaenaceae, and Sphaerosepalaceae), one to Seychelles (Mesdusagynaceae) and one to the Mascarene Islands (Psiloxylaceae). Twelve plant families are endemic or nearly endemic to South Africa (including Curtisiaceae, Heteropyxidaceae, Penaeaceae, Psiloxylaceae and Rhynchocalycaceae) of which five are endemic to the Cape floristic province (including Grubbiaceae). Other endemic Afrotropic families include Barbeyaceae, Montiniaceae, Myrothamnaceae and Oliniaceae.

Animals

The East African Great Lakes (Victoria, Malawi, and Tanganyika) are the center of biodiversity of many freshwater fishes, especially cichlids (they harbor more than two-thirds of the estimated 2,000 species in the family). The West African coastal rivers region covers only a fraction of West Africa, but harbours 322 of West African's fish species, with 247 restricted to this area and 129 restricted even to smaller ranges. The central rivers fauna comprises 194 fish species, with 119 endemics and only 33 restricted to small areas.

The Afrotropic has various endemic bird families, including ostriches (Struthionidae), sunbirds, the secretary bird (Sagittariidae), guineafowl (Numididae), and mousebirds (Coliidae). Also, several families of passerines are limited to the Afrotropics; These include rock-jumpers (Chaetopidae) and rockfowl (Picathartidae).

Africa has three endemic orders of mammals, the Tubulidentata (aardvarks), Afrosoricida (tenrecs and golden moles), and Macroscelidea (elephant shrews). The East-African plains are well known for their diversity of large mammals.

Four species of Great Apes (Hominidae) are endemic to Africa: both species of gorilla (western gorilla, *Gorilla gorilla*, and eastern gorilla, *Gorilla beringei*) and both species of chimpanzee (common chimpanzee, *Pan troglodytes*, and bonobo, *Pan paniscus*). Humans and their ancestors originated in Africa.

Ecoregions of Madagascar

The ecoregions of Madagascar, as defined by the World Wildlife Fund, include seven terrestrial, five freshwater, and two marine ecoregions. Madagascar's diverse natural habitats harbour a rich fauna and flora with high levels of endemism, but most ecoregions suffer from habitat loss.

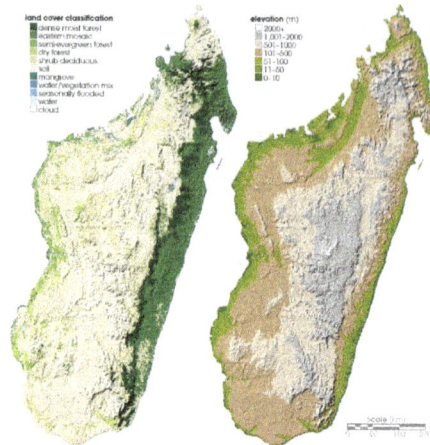

Land cover (left) and topography (right) of Madagascar.

Overview

Madagascar belongs to the Afrotropical realm. With its neighboring Indian Ocean islands, it has been classified by botanist Armen Takhtajan as *Madagascan Region*, and in phytogeography it is the floristic phytochorion *Madagascan Subkingdom* in the Paleotropical Kingdom. Madagascar features very contrasting topography, climate, and geology. A mountain range on the east, rising to 2,876 m (9,436 ft) at its highest point, captures most rainfall brought in by trade winds from the Indian Ocean. Consequently, the eastern belt harbours most of the humid forests, while precipitation decreases to the west. The rain shadow region in the southwest has a sub-arid climate. Temperatures are highest on the west coast, with annual means of up to 30 °C (86 °F), while the high massifs have a cool climate, with a 5 °C (41 °F) annual mean locally. Geology features mainly igneous and metamorphic basement rocks, with some lava and quartzite in the central and eastern plateaus, while the western part has belts of sandstone, limestone (including the tsingy formations), and unconsolidated sand.

Terrestrial Ecoregions

Seven terrestrial ecoregions are defined by the World Wildlife Fund for Madagascar. They range from the very humid eastern lowland forests to the sub-arid spiny thickets in the southwest.

Freshwater Ecoregions

Freshwater ecoregions correspond to major catchment areas with a distinctive assemblage of species. In Madagascar, five regions are distinguished:

- Eastern Lowlands

- Eastern Highlands

- Northwestern Basins

- Southern Basins

- Western Basins

Lake Ravelobe in Ankarafantsika National Park.

Marine Ecoregions

The seas around Madagascar are part of the Western Indian Ocean province in the Western Indo-Pacific realm. They are divided into two marine ecoregions:

- Southeast Madagascar

- Western and Northern Madagascar

Antarctic Realm

The Antarctic biogeographic realm

Antarctica is one of eight terrestrial biogeographic realms. The ecosystem includes Antarctica and several island groups in the southern Atlantic and Indian Oceans. The continent of Antarctica is so cold and dry that it has supported only 2 vascular plants for millions of years, and its flora presently consists of around 250 lichens, 100 mosses, 25-30 liverworts, and around 700 terrestrial and aquatic algal species, which live on the areas of exposed rock and soil around the shore of the continent. Antarctica's two flowering plant species, the Antarctic hair grass (*Deschampsia antarctica*) and Antarctic pearlwort (*Colobanthus quitensis*), are found on the northern and western parts of the Antarctic Peninsula. Antarctica is also home to a diversity of animal life, including penguins, seals, and whales.

Several Antarctic island groups are considered part of the Antarctica realm, including South Georgia and the South Sandwich Islands, South Orkney Islands, the South Shetland Islands, Bouvet Island, the Crozet Islands, Prince Edward Islands, Heard Island, the Kerguelen Islands, and the McDonald Islands. These islands have a somewhat milder climate than Antarctica proper, and support a greater diversity of tundra plants, although they are all too windy and cold to support trees.

Antarctic krill is the keystone species of the ecosystem of the Southern Ocean, and is an important food organism for whales, seals, leopard seals, fur seals, crabeater seals, squid, icefish, penguins, albatrosses and many other birds. The ocean there is so full of phytoplankton because around the ice continent water rises from the depths to the light flooded surface, bringing nutrients from all oceans back to the photic zone.

On August 20, 2014, scientists confirmed the existence of microorganisms living 800 metres (2,600 feet) below the ice of Antarctica.

History

Millions of years ago, Antarctica was warmer and wetter, and supported the Antarctic flora, including forests of podocarps and southern beech. Antarctica was also part of the ancient supercontinent of Gondwanaland, which gradually broke up by continental drift starting 110 million years ago. The separation of South America from Antarctica 30-35 million years ago allowed the Antarctic Circumpolar Current to form, which isolated Antarctica climatically and caused it to become much colder. The Antarctic flora subsequently died out in Antarctica, but is still an important component of the flora of southern Neotropic (South America) and Australasia, which were also former parts of Gondwana.

Some botanists recognize an Antarctic Floristic Kingdom that includes Antarctica, New Zealand, and parts of Temperate South America where the Antarctic Flora is still a major component.

Palearctic Realm

The Palearctic or Palaearctic is the largest of the eight biogeographic realms constituting the Earth's surface. It consists of Europe, Asia north of the foothills of the Himalayas, North Africa, and the northern and central parts of the Arabian Peninsula.

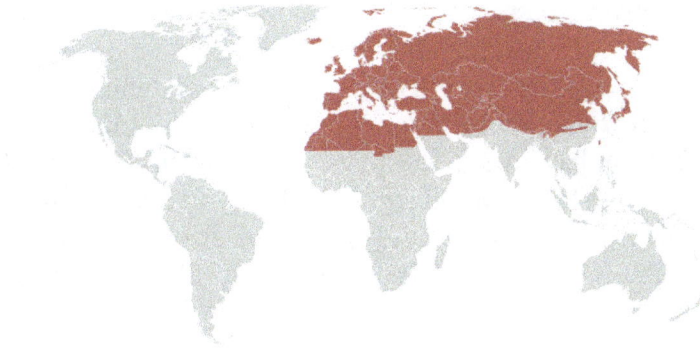

The Palearctic realm

The Palearctic realm comprises the smaller terrestrial ecoregions of the Euro-Siberian region; the Mediterranean Basin; the Sahara and Arabian Deserts; and Western, Central and East Asia. The Palaearctic realm also has numerous rivers and lakes, forming several freshwater ecoregions. Some of the rivers were the source of water for the earliest recorded civilizations that used irrigation methods.

Major ecological Regions

The Palearctic realm includes mostly boreal/subarctic-climate and temperate-climate ecoregions, which run across Eurasia from western Europe to the Bering Sea.

Euro-Siberian Region

The boreal and temperate Euro-Siberian region is the Palearctic's largest biogeographic region, which transitions from tundra in the northern reaches of Russia and Scandinavia to the vast taiga, the boreal coniferous forests which run across the continent. South of the taiga are a belt of temperate broadleaf and mixed forests and temperate coniferous forests. This vast Euro-Siberian region is characterized by many shared plant and animal species, and has many affinities with the temperate and boreal regions of the Nearctic ecoregion of North America. Eurasia and North America were often connected by the Bering land bridge, and have very similar mammal and bird fauna, with many Eurasian species having moved into North America, and fewer North American species having moved into Eurasia. Many zoologists consider the Palearctic and Nearctic to be a single Holarctic realm. The Palearctic and Nearctic also share many plant species, which botanists call the Arcto-Tertiary Geoflora.

Mediterranean Basin

The lands bordering the Mediterranean Sea in southern Europe, north Africa, and western Asia are home to the Mediterranean Basin ecoregions, which together constitute the world's largest and most diverse mediterranean climate region of the world, with generally mild, rainy winters and hot, dry summers. The Mediterranean basin's mosaic of Mediterranean forests, woodlands, and scrub are home to 13,000 endemic species. The Mediterranean basin is also one of the world's most endangered biogeographic regions; only 4% of the region's original vegetation remains, and human activities, including overgrazing, deforestation, and conversion of lands for pasture, agriculture, or urbanization, have degraded much of the region. Formerly the region was mostly covered with forests and woodlands, but heavy human use has reduced much of the region to the sclerophyll shrublands known as chaparral, matorral, maquis, or garrigue. Conservation International has designated the Mediterranean basin as one of the world's biodiversity hotspots.

Sahara and Arabian Deserts

A great belt of deserts, including the Atlantic coastal desert, Sahara desert, and Arabian desert, separates the Palearctic and Afrotropic ecoregions. This scheme includes these desert ecoregions in the palearctic realm; other biogeographers identify the realm boundary as the transition zone between the desert ecoregions and the Mediterranean basin ecoregions to the north, which places the deserts in the Afrotropic, while others place the boundary through the middle of the desert.

Western and Central Asia

The Caucasus mountains, which run between the Black Sea and the Caspian Sea, are a particularly rich mix of coniferous, broadleaf, and mixed forests, and include the temperate rain forests of the Euxine-Colchic deciduous forests ecoregion.

Central Asia and the Iranian plateau are home to dry steppe grasslands and desert basins, with montane forests, woodlands, and grasslands in the region's high mountains and plateaux. In southern Asia the boundary of the Palearctic is largely altitudinal. The middle altitude foothills of the Himalaya between about 2000–2500 m form the boundary between the Palearctic and Indomalaya ecoregions.

East Asia

China, Korea and Japan are more humid and temperate than adjacent Siberia and Central Asia, and are home to rich temperate coniferous, broadleaf, and mixed forests, which are now mostly limited to mountainous areas, as the densely populated lowlands and river basins have been converted to intensive agricultural and urban use. East Asia was not much affected by glaciation in the ice ages, and retained 96 percent of Pliocene tree genera, while Europe retained only 27 percent. In the subtropical region of southern China and southern edge of the Himalayas, the Palearctic temperate forests transition to the subtropical and tropical forests of Indomalaya, creating a rich and diverse mix of plant and animal species. The mountains of southwest China are also designated as a biodiversity hotspot. In Southeastern Asia, high mountain ranges form tongues of Palearctic flora and fauna in northern Indochina and southern China. Isolated small outposts (sky islands) occur as far south as central Myanmar (on Nat Ma Taung, 3050 m), northernmost Vietnam (on Fan Si Pan, 3140 m) and the high mountains of Taiwan.

Freshwater

The realm contains several important freshwater ecoregions as well, including the heavily developed rivers of Europe, the rivers of Russia, which flow into the Arctic, Baltic, Black, and Caspian seas, Siberia's Lake Baikal, the oldest and deepest lake on the planet, and Japan's ancient Lake Biwa.

Flora and Fauna

One bird family, the accentors (Prunellidae) is endemic to the Palearctic region. The Holarctic has four other endemic bird families: the divers or loons (Gaviidae), grouse (Tetraoninae), auks (Alcidae), and waxwings (Bombycillidae).

There are no endemic mammal orders in the region, but several families are endemic: Calomyscidae (mouse-like hamsters), Prolagidae, and Ailuridae (red pandas). Several mammal species originated in the Palearctic, and spread to the Nearctic during the Ice Age, including the brown bear (*Ursus arctos*, known in North America as the grizzly), red deer (*Cervus elaphus*) in Europe and the closely related elk (*Cervus canadensis*) in far eastern Siberia, American bison (*Bison bison*), and reindeer (*Rangifer tarandus*, known in North America as the caribou).

Megafaunal Extinctions

Several large Palearctic animals became extinct from the end of the Pleistocene into historic times, including the Irish elk (*Megaloceros giganteus*), aurochs (*Bos primigenius*) woolly rhinoceros (*Coelodonta antiquitatis*), woolly mammoth (*Mammuthus primigenius*), North African elephant (*Loxodonta africana pharaoensis*), Chinese elephant (*Elephas maximus rubridens*), cave bear (*Ursus spelaeus*), and European lion (*Panthera leo europaea*).

Mediterranean Basin

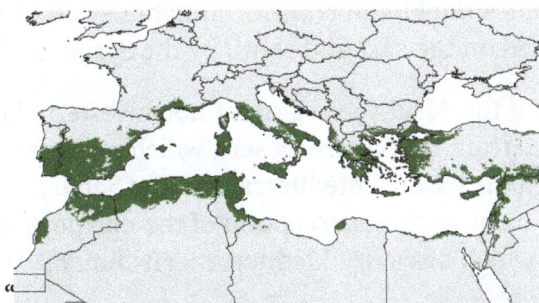

Potential distribution over the Mediterranean Basin of the olive tree—one of the best biological indicators of the Mediterranean Region (Oteros, 2014)

Political Map of the Mediterranean Basin

In biogeography, the Mediterranean Basin (also known as the Mediterranean region or sometimes Mediterranea) is the region of lands around the Mediterranean Sea that have a Mediterranean climate, with mild, rainy winters and hot, dry summers, which supports characteristic Mediterranean forests, woodlands, and scrub vegetation.

As a rule of thumb, the Mediterranean Basin is the Old World region where olive trees grow. However olive trees grow in other corners of the world which have a Mediterranean climate, and there are many areas around the Mediterranean Sea which do not have a Mediterranean climate and where olive trees cannot grow.

Geography

The Mediterranean basin covers portions of three continents Africa, Asia, and Europe.

It has a varied and contrasting topography. The Mediterranean Region offers an ever changing landscape of high mountains, rocky shores, impenetrable scrub, semi-arid steppes, coastal wetlands, sandy beaches and a myriad islands of various shapes and sizes dotted amidst the clear blue sea. Contrary to the classic sandy beach images portrayed in most tourist brochures, the Mediterranean is surprisingly hilly. Mountains can be seen from almost anywhere.

The Mediterranean Basin extends into Western Asia, covering the western and southern portions of the peninsula of Turkey, excluding the temperate-climate mountains of central Turkey. It includes the Mediterranean climate Levant at the eastern end of the Mediterranean, bounded on the east and south by the Syrian and Negev deserts.

The northern portion of the Maghreb region of northwestern Africa has a Mediterranean climate, separated from the Sahara Desert, which extends across North Africa, by the Atlas Mountains. In the eastern Mediterranean the Sahara extends to the southern shore of the Mediterranean, with the exception of the northern fringe of the peninsula of Cyrenaica in Libya, which has a dry Mediterranean climate.

Europe lies to the north, and three large Southern European peninsulas, the Iberian Peninsula, Italian Peninsula, and the Balkan Peninsula, extend into the Mediterranean-climate zone. A system of folded mountains, including the Pyrenees dividing Spain from France, the Alps dividing Italy from Central Europe, the Dinaric Alps along the eastern Adriatic, and the Balkan and Rhodope mountains of the Balkan Peninsula divide the Mediterranean from the temperate climate regions of Western and Central Europe.

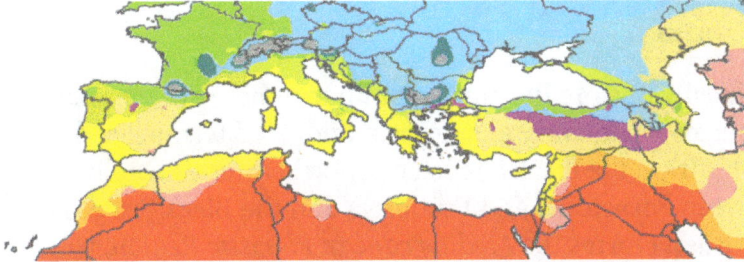

Köppen-Geiger-based map of the areas surrounding the Mediterranean Sea. Based on the work of M. C. Peel, B.L. Finlayson and T.A. McMahon at the University of Melbourne.

Geology and Paleoclimatology

The Mediterranean Basin was shaped by the ancient collision of the northward-moving African-Arabian continent with the stable Eurasian continent. As Africa-Arabia moved north, it closed the former Tethys Sea, which formerly separated Eurasia from the ancient super continent of Gondwana, of which Africa was part. At about the same time, 170 mya in the Jurassic period, a small Neotethys ocean basin formed shortly before the Tethys Sea was closed at the eastern end. The collision pushed up a vast system of mountains, extending from the Pyrenees in Spain to the Zagros Mountains in Iran. This episode of mountain building, known as the Alpine orogeny, occurred mostly during the Oligocene (34 to 23 million years ago (mya)) and Miocene (23 to 5.3 mya) epochs. The Neotethys became larger during these collisions and associated folding and subduction.

About 6 mya during the late Miocene, the Mediterranean was closed at its western end by drifting Africa, which caused the entire sea to evaporate. There followed several (debated) episodes of sea drawdown and re-flooding known as the Messinian Salinity Crisis, which ended when the Atlantic last re-flooded the basin at the end of the Miocene. Recent research has suggested that a desiccation-flooding cycle may have repeated several times during the last 630,000 years of the Miocene epoch, which could explain several events of large amounts of salt deposition. Recent studies, however, show that repeated desiccation and re-flooding is unlikely from a geodynamic point of view.

The end of the Miocene also marked a change in the Mediterranean Basin's climate. Fossil evidence shows that the Mediterranean Basin had a relatively humid subtropical climate with summer rainfall during the Miocene, which supported laurel forests. The shift to a Mediterranean climate occurred within the last 3.2–2.8 million years, during the Pliocene epoch, as summer rainfall decreased. The subtropical laurel forests retreated, although they persisted on the islands of Macaronesia off the Atlantic coast of Iberia and North Africa, and the present Mediterranean vegetation evolved, dominated by coniferous trees and sclerophyllous trees and shrubs, with small, hard, waxy leaves that prevent moisture loss in the dry summers. Much of these forests and shrublands have been altered beyond recognition by thousands of years of human habitation. There are now very few relatively intact natural areas in what was once a heavily wooded region.

Flora and Fauna

Phytogeographically, the Mediterranean basin together with the nearby Atlantic coast, the Mediterranean woodlands and forests and Mediterranean dry woodlands and steppe of North Africa, the Black Sea coast of northeastern Anatolia, the southern coast of Crimea between Sevastopol and Feodosiya and the Black Sea coast between Anapa and Tuapse in Russia forms the *Mediterranean Floristic Region*, which belongs to the Tethyan Subkingdom of the Boreal Kingdom and is enclosed between the Circumboreal, Irano-Turanian, Saharo-Arabian and Macaronesian floristic regions.

The Mediterranean Region was first proposed by German botanist August Grisebach in the late 19th century.

Drosophyllaceae, recently segregated from Droseraceae, is the only plant family endemic to the region. Among the endemic plant genera are:

- *Tetraclinis*
- *Rupicapnos*
- *Ceratocapnos*
- *Soleirolia*
- *Ortegia*
- *Bolanthus*
- *Lycocarpus*
- *Ionopsidium*
- *Bivonaea*
- *Euzomodendron*
- *Hutera*
- *Vella*
- *Boleum*
- *Didesmus*
- *Morisia*
- *Guiraoa*
- *Malope*

- *Drosophyllum*
- *Ceratonia*
- *Chronanthus*
- *Anagyris*
- *Callicotome*
- *Spartium*
- *Hymenocarpus*
- *Biserrula*
- *Argania*
- *Petagnia*
- *Lagoecia*
- *Putoria*
- *Fedia*
- *Tremastelma*
- *Bellardia*
- *Lafuentea*
- *Rosmarinus*

- *Argantoniella*
- *Preslia*
- *Gyrocarion*
- *Dorystoechas*
- *Coridothymus*
- *Trachelium*
- *Santolina*
- *Cladanthus*
- *Staehelina*
- *Leuzea*
- *Andryala*
- *Rothmaleria*
- *Chionodoxa*
- *Hermodactylus*
- *Triplachne*
- *Helicodiceros*
- *Chamaerops*

The genera *Aubrieta*, *Sesamoides*, *Cynara*, *Dracunculus*, *Arisarum* and *Biarum* are nearly endemic. Among the endemic species prominent in the Mediterranean vegetation are the Aleppo pine, stone pine, Mediterranean cypress, bay laurel, Oriental sweetgum, holm oak, kermes oak, strawberry tree, Greek strawberry tree, mastic, terebinth, common myrtle, oleander, *Acanthus mollis* and *Vitex agnus-castus*. Moreover, many plant taxa are shared with one of the four neighboring floristic regions only. According to different versions of Armen Takhtajan's delineation, the Mediterranean Region is further subdivided into seven to nine floristic provinces: Southwestern Mediterranean (or Southern Moroccan and Southwestern Mediterranean), Ibero-Balearian (or Iberian and Balearian), Liguro-Tyrrhenian, Adriatic, East Mediterranean, South Mediterranean and Crimeo-Novorossiysk.

The Mediterranean Basin is the largest of the world's five Mediterranean forests, woodlands, and scrub regions. It is home to a number of plant communities, which vary with rainfall, elevation, latitude, and soils.

- Scrublands occur in the driest areas, especially areas near the seacoast where wind and salt spray are frequent. Low, soft-leaved scrublands around the Mediterranean are known as *garrigar* in Catalan, *garrigue* in French, *phrygana* in Greek, *tomillares* in Spanish, and *batha* in Hebrew.

- Shrublands are dense thickets of evergreen sclerophyll shrubs and small trees, and are the most common plant community around the Mediterranean. Mediterranean shrublands are known as *màquia* in Catalan, *macchia* in Italian, *maquis* in French, and "matorral" in Spanish. In some places shrublands are the mature vegetation type, and in other places the result of degradation of former forest or woodland by logging or overgrazing, or disturbance by major fires.

- Savannas and grasslands occur around the Mediterranean, usually dominated by annual grasses.

- Woodlands are usually dominated by oak and pine, mixed with other sclerophyll and coniferous trees.

- Forests are distinct from woodlands in having a closed canopy, and occur in the areas of highest rainfall and in riparian zones along rivers and streams where they receive summer water. Mediterranean forests are generally composed of evergreen trees, predominantly oak and pine. At higher elevations Mediterranean forests transition to mixed broadleaf and tall conifer forests similar to temperate zone forests.

The Mediterranean Basin is home to considerable biodiversity, including 22,500 endemic vascular plant species. Conservation International designates the region as a biodiversity hotspot, because of its rich biodiversity and its threatened status. The Mediterranean Basin has an area of 2,085,292 km², of which only 98,009 km² remains undisturbed.

Endangered mammals of the Mediterranean Basin include the Mediterranean monk seal, the Barbary macaque, and the Iberian lynx.

Ecoregions

- Aegean and Western Turkey sclerophyllous and mixed forests (Greece, Turkey)

- Anatolian conifer and deciduous mixed forests (Turkey)

- Canary Islands dry woodlands and forests (Spain)

- Corsican montane broadleaf and mixed forests (France)

- Crete Mediterranean forests (Greece)

- Cyprus Mediterranean forests (Cyprus)

- Eastern Mediterranean conifer-sclerophyllous-broadleaf forests (Lebanon, Israel, the West Bank, the Gaza Strip, Jordan, Syria, Turkey)

- Iberian conifer forests (Portugal, Spain)

- Iberian sclerophyllous and semi-deciduous forests (Portugal, Spain)

- Illyrian deciduous forests (Albania, Bosnia and Herzegovina, Croatia, Greece, Italy, Slovenia)

- Italian sclerophyllous and semi-deciduous forests (France, Italy)

- Mediterranean acacia-argania dry woodlands and succulent thickets (Morocco, Canary Islands (Spain))

- Mediterranean dry woodlands and steppe (Algeria, Egypt, Libya, Morocco, Tunisia)

- Mediterranean woodlands and forests (Algeria, Morocco, Tunisia)

- Northeastern Spain and Southern France Mediterranean forests (France, Spain)

- Northwest Iberian montane forests (Portugal, Spain)

- Pindus Mountains mixed forests (Albania, Greece, Macedonia)

- South Apennine mixed montane forests (Italy)

- Southeastern Iberian shrubs and woodlands (Spain)

- Southern Anatolian montane conifer and deciduous forests (Lebanon, Israel, Jordan, Syria, Turkey)

- Southwest Iberian Mediterranean sclerophyllous and mixed forests (France, Italy, Morocco, Portugal, Spain)

- Tyrrhenian-Adriatic sclerophyllous and mixed forests (Croatia, France, Italy, Malta)

History

Neanderthals inhabited western Asia and the non-glaciated portions of Europe starting about 230,000 years ago. Modern humans moved into western Asia from Africa less than 100,000 years ago. Modern humans, known as Cro-Magnons, moved into Europe approximately 50-40,000 years ago.

The most recent glacial period, the Wisconsin glaciation, reached its maximum extent approximately 21,000 years ago, and ended approximately 12,000 years ago. A warm period, known as the Holocene climatic optimum, followed the ice age.

Food crops, including wheat, chickpeas, and olives, along with sheep and goats, were domesticated in the eastern Mediterranean in the 9th millennium BCE, which allowed for the establishment of agricultural settlements. Near Eastern crops spread to southeastern Europe in the 7th millennium BCE. Poppy and oats were domesticated in Europe from the 6th to the 3rd millennium BCE. Agricultural settlements spread around the Mediterranean Basin. Megaliths were constructed in Europe from 4500 – 1500 BCE.

A strengthening of the summer monsoon 9000–7000 years ago increased rainfall across the Sahara, which became a grassland, with lakes, rivers, and wetlands. After a period of climatic instability, the Sahara settled into a desert state by the 4th millennium BCE.

Agriculture

Wheat is the dominant grain grown around the Mediterranean Basin. Pulses and vegetables are also grown. The characteristic tree crop is the olive. Figs are another important fruit tree, and citrus, especially lemons, are grown where irrigation is present. Grapes are an important vine crop, grown for fruit and to make wine. Rice and summer vegetables are grown in irrigated areas.

Nearctic Realm

The Nearctic is one of the eight biogeographic realms constituting the Earth's land surface.

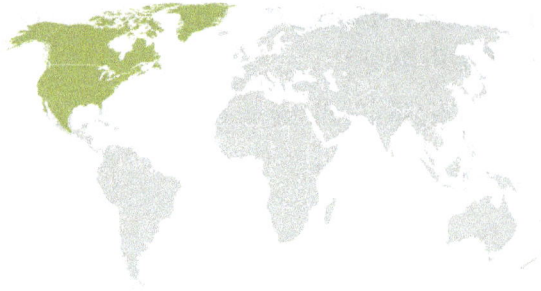

The Nearctic realm

The Nearctic realm covers most of North America, including Greenland, Central Florida, and the highlands of Mexico. The parts of North America that are not in the Nearctic realm are Eastern Mexico, Southern Florida, Central America, and the Caribbean islands which are part of the Neotropical realm, together with South America.

Major Ecological Regions

The World Wildlife Fund (WWF) divides the Nearctic into four bioregions, defined as "geographic clusters of ecoregions that may span several habitat types, but have strong biogeographic affinities, particularly at taxonomic levels higher than the species level (genus, family)."

Canadian Shield

The Canadian Shield bioregion extends across the northern portion of the continent, from the Aleutian Islands to Newfoundland. It includes the Nearctic's Arctic Tundra and Boreal forest ecoregions.

In terms of floristic provinces, it is represented by part of the Canadian Province of the Circumboreal Region.

Eastern North America

The Eastern North America bioregion includes the temperate broadleaf and mixed forests of the Eastern United States and southeastern Canada, the Great Plains temperate grasslands of the Central United States and south-central Canada, the temperate coniferous forests of the Southeastern United States and Central Florida. In terms of floristic provinces, it is represented by the North American Atlantic Region and part of the Canadian Province of the Circumboreal Region.

Western North America

The Western North America bioregion includes the temperate coniferous forests of the coastal and mountain regions of southern Alaska, western Canada, and the Western United States from the Pacific Coast and Northern California to the Rocky Mountains, as well as the cold-winter intermountain deserts and xeric shrublands and temperate grasslands and shrublands of the Western United States.

In terms of floristic provinces, it is represented by the Rocky Mountain region.

Northern Mexico and Southwestern North America

The Northern Mexico bioregion includes the mild-winter to cold-winter deserts and xeric shrublands of northern Mexico, Southern California, and the Southwestern United States, including the Chihuahuan, Sonoran, and Mojave Deserts. The Mediterranean climate ecoregions of the Southern and Central Coast of California include the California chaparral and woodlands, California coastal sage and chaparral, California interior chaparral and woodlands, and California montane chaparral and woodlands.

The bioregion also includes the warm temperate and subtropical pine and pine-oak forests, including the Arizona Mountains forests and the Sierra Madre Occidental, Sierra Madre Oriental, and Sierra Juarez and San Pedro Martir pine-oak forests.

In terms of floristic provinces, it is represented by the Madrean Region.

History

Although North America and South America are presently joined by the Isthmus of Panama, these continents were separated for about 180 million years, and evolved very different plant and animal lineages. When the ancient supercontinent of Pangaea split into two about 180 million years ago, North America remained joined to Eurasia as part of the supercontinent of Laurasia, while South America was part of the supercontinent of Gondwana. North America later split from Eurasia. North America has been joined by land bridges to both Asia and South America since then, which allowed an exchange of plant and animal species between the continents, the Great American Interchange.

A former land bridge across the Bering Strait between Asia and North America allowed many plants and animals to move between these continents, and the Nearctic realm shares many plants and animals with the Palearctic. The two realms are sometimes included in a single Holarctic realm.

Many large animals, or megafauna, including horses, camels, mammoths, mastodonts, ground sloths, sabre-tooth cats (*Smilodon*), the giant short-faced bear (*Arctodus simus*), and the cheetah, became extinct in North America at the end of the Pleistocene epoch (ice ages), at the same time the first evidence of humans appeared, in what is called the Holocene extinction event. Previously, the megafaunal extinctions were believed to have been caused by the changing climate, but many scientists now believe, while the climate change contributed to these extinctions, the primary cause was hunting by newly arrived humans or, in the case of some large predators, extinction resulting from prey becoming scarce. The American bison (*Bison bison*), brown bear or grizzly bear (*Ursus arctos*), moose or Eurasian elk (*Alces alces*), and elk or wapiti (*Cervus canadensis*) entered North America around the same time as the first humans, and expanded rapidly, filling ecological niches left empty by the newly extinct North American megafauna.

Flora and Fauna

Flora and Fauna that Originated in the Nearctic

Animals originally unique to the Nearctic include:

- Family Canidae, dogs, wolves, foxes, and coyotes

- Family Camelidae, camels and their South American relatives including the llama. now extinct in the Nearctic

- Family Equidae, horses, donkeys and their relatives. now only found in the Nearctic as feral horses

- Family Tapiridae, tapirs now extinct in the Nearctic

- Family Antilocapridae, last survivor of which is the pronghorn

- Subfamily Tremarctinae, or short-faced, bears, including the giant short-faced bear (Arctodus simius) One other member of the group is the spectacled bear (Tremarctos ornatus) of South America. now extinct in the Nearctic

- The American cheetah (Miracinonyx) now extinct worldwide

Flora and Fauna Endemic to the Nearctic

One bird family, the wrentits (Timaliinae), is endemic to the Nearctic region. The Holarctic has four endemic families: divers (Gaviidae), grouse (Tetraoninae), auks (Alcidae), and the waxwings (Bombycillidae). The scarab beetle families Pleocomidae and Diphyllostomatidae (Coleoptera) are also endemic to the Nearctic. The fly species *Cynomya cadaverina* is also found in high numbers in this area.

Plants families endemic or nearly endemic to the Nearctic include the Crossosomataceae, Simmondsiaceae, and Limnanthaceae.

Neotropical Realm

The Neotropical realm is one of the eight biogeographic realms constituting the Earth's land surface. Physically, it includes the tropical terrestrial ecoregions of the Americas and the entire South American temperate zone.

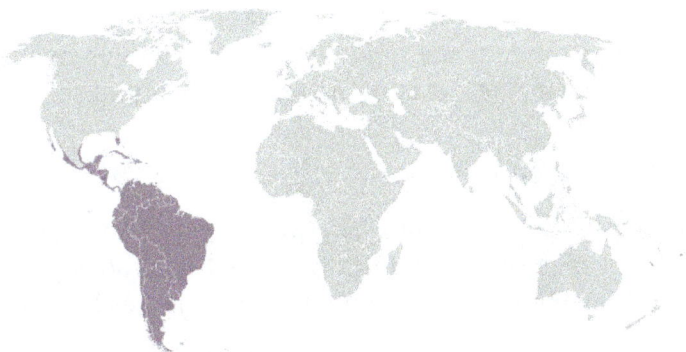

The Neotropical realm

Definition

In biogeography, the Neotropic or Neotropical realm is one of the eight terrestrial realms. This realm includes South and Central America, and in North America the southern Mexican lowlands, the Caribbean islands, and southern Florida, because these regions share a large number of plant and animal groups.

The realm also includes temperate southern South America. In contrast, the Neotropical Floristic Kingdom excludes southernmost South America, which instead is placed in the Antarctic kingdom.

The Neotropic is delimited by similarities in fauna or flora. Its fauna and flora are distinct from the Nearctic (which includes most of North America) because of the long separation of the two continents. The formation of the Isthmus of Panama joined the two continents two to three million years ago, precipitating the Great American Interchange, an important biogeographical event.

The Neotropic includes more tropical rainforest (tropical and subtropical moist broadleaf forests) than any other realm, extending from southern Mexico through Central America and northern South America to southern Brazil, including the vast Amazon Rainforest. These rainforest ecoregions are one of the most important reserves of biodiversity on Earth. These rainforests are also home to a diverse array of indigenous peoples, who to varying degrees persist in their autonomous and traditional cultures and subsistence within this environment. The number of these peoples who are as yet relatively untouched by external influences continues to decline significantly, however, along with the near-exponential expansion of urbanization, roads, pastoralism and forest industries which encroach on their customary lands and environment. Nevertheless, amidst these declining circumstances this vast "reservoir" of human diversity continues to survive, albeit much depleted. In South America alone, some 350–400 indigenous languages and dialects are still living (down from an estimated 1,500 at the time of first European contact), in about 37 distinct language families and a further number of unclassified and isolate languages. Many of these languages and their cultures are also endangered. Accordingly, conservation in the Neotropical realm is a hot political concern, and raises many arguments about development versus indigenous versus ecological rights and access to or ownership of natural resources.

Major Ecological Regions

The WWF subdivides the realm into *bioregions*, defined as "geographic clusters of ecoregions that may span several habitat types, but have strong biogeographic affinities, particularly at taxonomic levels higher than the species level (genus, family)."

Laurel forest and other cloud forest are subtropical and mild temperate forest, found in areas with high humidity and relatively stable and mild temperatures. Tropical rainforest, tropical and subtropical moist broadleaf forests are highlight in Southern North America, Amazonia, Caribbean, Central America, Northern Andes and Central Andes.

Amazonia

The Amazonia bioregion is mostly covered by tropical moist broadleaf forest, including the vast Amazon rainforest, which stretches from the Andes mountains to the Atlantic Ocean, and the lowland forests of the Guianas. The bioregion also includes tropical savanna and tropical dry forest ecoregions.

Central Andes

Eastern South America

Eastern South America includes the Caatinga xeric shrublands of northeastern Brazil, the broad Cerrado grasslands and savannas of the Brazilian Plateau, and the Pantanal and Chaco grasslands. The diverse Atlantic forests of eastern Brazil are separated from the forests of Amazonia by the Caatinga and Cerrado, and are home to a distinct flora and fauna.

Northern Andes

Orinoco

The Orinoco is a region of humid forested broadleaf forest and wetland primarily comprising the drainage basin for the Orinoco River and other adjacent lowland forested areas. This region includes most of Venezuela and parts of Colombia.

Southern South America

The temperate forest ecoregions of southwestern South America, including the temperate rain forests of the Valdivian temperate rain forests and Magellanic subpolar forests ecoregions, and the Juan Fernández Islands and Desventuradas Islands, are a refuge for the ancient Antarctic flora, which includes trees like the southern beech (*Nothofagus*), podocarps, the alerce (*Fitzroya cupressoides*), and Araucaria pines like the monkey-puzzle tree (*Araucaria araucana*). These magnificent rainforests are endangered by extensive logging and their replacement by fast-growing non-native pines and eucalyptus.

History

South America was originally part of the supercontinent of Gondwana, which included Africa, Australia, India, New Zealand, and Antarctica, and the Neotropic shares many plant and animal lineages with these other continents, including marsupial mammals and the Antarctic flora.

After the final breakup of the Gondwana about 110 million years ago, South America was separated from Africa and drifted north and west. Much later, about two to three million years ago, South America was joined with North America by the formation of the Isthmus of Panama, which allowed a biotic exchange between the two continents, the Great American Interchange. South American species like the ancestors of the Virginia opossum (*Didelphis virginiana*) and the armadillo moved into North America, and North Americans like the ancestors of South America's camelids, including the llama (*Lama glama*), moved south. The long-term effect of the exchange was the extinction of many South American species, mostly by outcompetition by northern species.

Endemic Animals and Plants

Animals

31 bird families are endemic to the Neotropical realm, over twice the number of any other realm. They include tanagers, rheas, tinamous, curassows, antbirds, ovenbirds, and toucans. Bird families originally unique to the Neotropics include hummingbirds (family Trochilidae) and wrens (family Troglodytidae).

Mammal groups originally unique to the Neotropics include:

- Order *Xenarthra*: anteaters, sloths, and armadillos

- New World monkeys

- Caviomorpha rodents, including capybaras and guinea pigs, and chinchillas

- American opossums (order *Didelphimorphia*) and shrew opossums (order *Paucituberculata*)

43 fish families and subfamilies are endemic to the Neotropical realm, more than any other realm (Reis et al., 2003). Neotropical fishes include more than 5,700 species,

and represent at least 66 distinct lineages in continental freshwaters (Albert and Reis, 2011). The well-known red-bellied piranha is endemic to the Neotropic realm, occupying a larger geographic area than any other piranha species. Some fish groups originally unique to the Neotropics include:

- Order Gymnotiformes: Neotropical electric fishes
- Family Characidae: tetras and allies
- Family Loricariidae: armoured catfishes
- Subfamily Cichlinae: Neotropical cichlids
- Subfamily Poeciliinae: guppies and relatives

Examples of other animal groups that are entirely or mainly restricted to the Neotropical region include:

- Caimans
- New World coral snakes
- Poison dart frogs
- Dactyloidae ("anoles")
- Preponini and Anaeini butterflies (including *Agrias*)
- Brassolini and Morphini butterflies (including *Caligo* and *Morpho*)
- Callicorini butterflies
- Heliconiini butterflies
- Firetips or firetail skipper butterflies
- Euglossini bees
- Augochlorini bees
- Pseudostigmatidae ("giant damselflies")
- Mantoididae (short-bodied mantises)
- Canopidae, Megarididae, and Phloeidae (pentatomoid bugs)
- Aetalionidae and Melizoderidae (treehoppers)

Plants

Plant families endemic and partly subendemic to the realm are, according to Takhtajan (1978), Hymenophyllopsidaceae, Marcgraviaceae, Caryocaraceae, Pellicieraceae, Quiinaceae, Peridiscaceae, Bixaceae, Cochlospermaceae, Tovariaceae, Lissocarpaceae (*Lissocarpa*), Brunelliaceae, Dulongiaceae, Columelliaceae, Julianiaceae, Picro-dendraceae, Goupiaceae, Desfontainiaceae, Plocospermataceae, Dialypetalanthaceae (*Dialypetalanthus*), Nolanaceae (*Nolana*), Calyceraceae, Heliconiaceae, Cannaceae, Thurniaceae and Cyclanthaceae.

Plant families that originated in the Neotropic include Bromeliaceae, Cannaceae and Heliconiaceae.

Plant species with economic importance originally unique to the Neotropic include:

- Potato (*Solanum tuberosum*)

- Tomato (*Solanum lycopersicum*)

- Cacao tree (*Theobroma cacao*), source of cocoa and chocolate

- Maize (*Zea mays*)

- Lima bean (*Phaseolus lunatus*)

- Cotton (*Gossypium barbadense*)

- Cassava (*Manihot esculenta*)

- Sweet potato (*Ipomoea batatas*)

- Amaranth (*Amaranthus caudatus*)

- Quinoa (*Chenopodium quinoa*)

Caribbean Bioregion

The Caribbean bioregion is a biogeographic region that includes the islands of the Caribbean Sea, which share a fauna, flora and mycobiota distinct from surrounding bioregions.

The Caribbean bioregion, as described by the World Wildlife Fund, includes the Greater Antilles (Cuba, Hispaniola, Puerto Rico, and Jamaica), the Lesser Antilles, the Bahamas and Turks and Caicos Islands, and Aruba, Bonaire, and Curaçao. It does not include Trinidad and Tobago; these islands rest on South America's shallow continental shelf, and have been historically part of the South American continent.

The climate of the ecoregion is tropical, and varies from humid to arid. Geology and topography also vary, with larger mountainous islands of continental rock, volcanic islands, and low-lying coral and limestone islands. The bioregion includes tropical moist forests, tropical dry forests, tropical pine forests, flooded grasslands and savannas, xeric shrublands, and mangroves.

The Caribbean bioregion's distinct fauna, flora and mycobiota was shaped by long periods of physical separation from the neighboring continents, allowing animals, fungi and plants to evolve in isolation. Other animals, fungi and plants arrived via long-distance oceanic dispersal or island hopping from North America and South America.

Three mammal families are endemic to the bioregion; the Solenodontidae includes two species of *Solenodon*, one species on Cuba, the other on Hispaniola. Fossil evidence shows that the family was once more widespread in North America. Family Nesophontidae, or the West Indian shrews, contained a single genus, *Nesophontes*, which inhabited Cuba, Hispaniola, Puerto Rico and the Cayman Islands. All members of the family are now believed to be extinct. The Capromyidae, or hutias, include a number of species, mainly from the Greater Antilles. Many other rodents of the Caribbean are also restricted to the region.

Ecoregions

Tropical and subtropical moist broadleaf forests

- Cuban moist forests (Cuba)

- Hispaniolan moist forests (Dominican Republic, Haiti)

- Jamaican moist forests (Jamaica)

- Leeward Islands moist forests (Antigua, British Virgin Islands, Guadeloupe, Montserrat, Nevis, Saint Kitts, US Virgin Islands)

- Puerto Rican moist forests (Puerto Rico)

- South Florida rocklands (United States)

- Windward Islands moist forests (Dominica, Grenada, Martinique, Saint Lucia, Saint Vincent and the Grenadines)

Tropical and Subtropical Dry Broadleaf Forests

- Bahamian dry forests (Bahamas)

- Cayman Islands dry forests (Cayman Islands)

- Cuban dry forests (Cuba)

- Hispaniolan dry forests (Dominican Republic, Haiti)

- Jamaican dry forests (Jamaica)

- Leeward Islands dry forests (Anguilla, Antigua and Barbuda, Montserrat, Netherlands Antilles)

- Puerto Rican dry forests (Puerto Rico)

- Windward Islands dry forests (Grenada, Martinique, Saint Lucia, Saint Vincent and the Grenadines)

Tropical and Subtropical Coniferous Forests

- Bahamian pineyards (The Bahamas)
- Cuban pine forests (Cuba)
- Hispaniolan pine forests (Dominican Republic, Haiti)

Flooded Grasslands and Savannas

- Cuban wetlands (Cuba)
- Enriquillo wetlands (Dominican Republic, Haiti)
- Everglades (United States)

Deserts and Xeric Shrublands

- Aruba-Curaçao-Bonaire cactus scrub (Aruba, Bonaire, Curaçao)
- Cayman Islands xeric scrub (Cayman Islands)
- Cuban cactus scrub (Cuba)
- Leeward Islands xeric scrub (Anguilla, Antigua and Barbuda, British Virgin Islands, Guadeloupe, Saint Martin, Saint Barthelemy, Saba, US Virgin Islands)
- Windward Islands xeric scrub (Barbados, Dominica, Grenada, Martinique, Saint Lucia, Saint Vincent and the Grenadines)

Mangrove

- Bahamian mangroves (Bahamas, Turks and Caicos Islands)
- Greater Antilles mangroves (Cuba, Dominican Republic, Haiti, Jamaica, Puerto Rico)
- Lesser Antilles mangroves (Lesser Antilles)

Central America Bioregion

The Central America bioregion is a biogeographic region comprising southern Mexico and Central America.

The bioregion covers the southern portion of Mexico, all of Belize, Costa Rica, El Salvador, Guatemala, Honduras, and Nicaragua, and all but easternmost Panama.

WWF defines bioregions as "geographic clusters of ecoregions that may span several habitat types, but have strong biogeographic affinities, particularly at taxonomic levels higher than the species level (genus, family)."

The bioregion lies in the tropics, and is home to tropical moist broadleaf forests, tropical dry broadleaf forests, and tropical coniferous forests. The higher mountains are home to cool-climate montane forests, grasslands and shrublands.

Central America connects North America to South America. The land bridge was completed 2.8 million years ago, when the Isthmus of Panama was formed, linking the two continents for the first time in tens of millions of years. The resulting Great American Interchange of animals and plants shaped the flora and fauna of the Central America bioregion.

Large mammals include the white-lipped peccary *(Tayassu pecari)*, Baird's tapir *(Tapirus bairdii)*, white-tailed deer *(Odocoileus virginianus)*, Central American red brocket *(Mazama temama)*, Yucatan brown brocket *(Mazama pandora)*, giant anteater *(Myrmecophaga tridactyla)*, brown-throated sloth *(Bradypus variegatus)*, jaguar *(Panthera onca)*, cougar *(Puma concolor)*, and ocelot *(Leopardus pardalis)*.

Plants of South American origin came to dominate the tropical lowlands of Central America, as did South American freshwater fish and invertebrates. 95% of Central American freshwater fish are South American in origin, with only the Tropical gar *(Atractosteus tropicus)*, three clupeids *(Dorosoma)*, a catostomid *(Ictiobus)*, and an ictalurid *(Ictalurus)* of North American origin.

The montane vegetation of the region is distinct from the lowland vegetation, and includes species with origins in temperate North America, including oaks *(Quercus)*, Pines *(Pinus)* and alders *(Alnus)*, as well as a some species with origins in temperate South America, including *Weinmannia* and *Drimys*.

Ecoregions

Tropical and Subtropical Moist Broadleaf Forests

- Cayos Miskitos-San Andrés and Providencia moist forests (Colombia, Nicaragua)

- Central American Atlantic moist forests (Costa Rica, Nicaragua, Panama)

- Central American montane forests (El Salvador, Guatemala, Honduras, Mexico, Nicaragua)

- Chiapas montane forests (Mexico)

- Chimalapas montane forests (Mexico)

- Cocos Island moist forests (Costa Rica)

- Costa Rican seasonal moist forests (Costa Rica, Nicaragua)

- Eastern Panamanian montane forests (Colombia, Panama)

- Isthmian-Atlantic moist forests (Costa Rica, Nicaragua, Panama)

- Isthmian-Pacific moist forests (Costa Rica, Panama)

- Oaxacan montane forests (Mexico)

- Pantanos de Centla (Mexico)

- Petén-Veracruz moist forests (Mexico)

- Sierra de los Tuxtlas (Mexico)

- Sierra Madre de Chiapas moist forest (El Salvador, Guatemala, Mexico)

- Talamancan montane forests (Costa Rica, Panama)

- Veracruz moist forest (Mexico)

- Veracruz montane forests (Mexico)

- Yucatán moist forests (Belize, Guatemala, Mexico)

Tropical and Subtropical Dry Broadleaf Forests

- Bajío dry forests (Mexico)

- Balsas dry forests (Mexico)

- Central American dry forests (Costa Rica, El Salvador, Guatemala, Honduras, Mexico, Nicaragua)

- Chiapas Depression dry forests (Guatemala, Mexico)

- Jalisco dry forests (Mexico)

- Panamanian dry forests (Panama)

- Revillagigedo Islands dry forests (Mexico)

- Sierra de la Laguna dry forests (Mexico)

- Sinaloan dry forests (Mexico)

- Southern Pacific dry forests (Mexico)

- Veracruz dry forests (Mexico)

- Yucatán dry forests (Mexico)

Tropical and Subtropical Coniferous Forests

- Belizean pine forests (Belize)
- Central American pine-oak forests (El Salvador, Guatemala, Honduras, Mexico, Nicaragua)
- Miskito pine forests (Honduras, Nicaragua)
- Sierra de la Laguna pine-oak forests (Mexico)
- Sierra Madre de Oaxaca pine-oak forests (Mexico)
- Sierra Madre del Sur pine-oak forests (Mexico)
- Trans-Mexican Volcanic Belt pine-oak forests (Mexico)

Tropical and Subtropical Grasslands, Savannas, and Shrublands

- Clipperton Island shrub and grasslands (Clipperton Island is an overseas territory of France)

Flooded Grasslands and Savannas

- Central Mexican wetlands (Mexico)
- Jalisco palm savannas (Mexico)

Montane Grasslands and Shrublands

- Talamancan páramo (Costa Rica)
- Zacatonal (Mexico)

Deserts and Xeric Shrublands

- Motagua Valley thornscrub (Guatemala)
- San Lucan xeric scrub (Mexico)
- Tehuacán Valley matorral (Mexico)

Mangrove

- Alvarado mangroves (Mexico)
- Belizean Coast mangroves (Belize)
- Belizean Reef mangroves (Belize)
- Bocas del Toro-San Bastimentos Island-San Blas mangroves (Costa Rica, Panama)

- Gulf of Fonseca mangroves (El Salvador, Honduras, Nicaragua)

- Gulf of Panama mangroves (Panama)

- Marismas Nacionales-San Blas mangroves (Mexico)

- Mayan Corridor mangroves (Mexico)

- Mexican South Pacific Coast mangroves (Mexico)

- Moist Pacific Coast mangroves (Costa Rica, Panama)

- Mosquitia-Nicaraguan Caribbean Coast mangroves (Costa Rica, Honduras, Nicaragua)

- Northern Dry Pacific Coast mangroves (El Salvador, Guatemala)

- Northern Honduras mangroves (Guatemala, Honduras)

- Petenes mangroves (Mexico)

- Ría Lagartos mangroves (Mexico)

- Rio Negro-Rio San Sun mangroves (Costa Rica, Nicaragua)

- Southern Dry Pacific Coast mangroves (Costa Rica, Nicaragua)

- Tehuantepec-El Manchon mangroves (Mexico)

- Usumacinta mangroves (Mexico)

Indomalayan Realm

The Indomalayan realm is one of the eight biogeographic realms. It extends across most of South and Southeast Asia and into the southern parts of East Asia.

The Indomalayan realm

Also called the Oriental realm by biogeographers, Indomalaya extends from Afghanistan through the Indian subcontinent and Southeast Asia to lowland southern China, and through Indonesia as far as Java, Bali, and Borneo, east of which lies the Wallace line, the realm boundary named after Alfred Russel Wallace which separates Indomalayan from Australasia. Indomalaya also includes the Philippines, lowland Taiwan, and Japan's Ryukyu Islands.

Most of Indomalaya was originally covered by forest, mostly tropical and subtropical moist broadleaf forests, with tropical and subtropical dry broadleaf forests predominant in much of India and parts of Southeast Asia. The tropical moist forests of Indomalaya are mostly dominated by trees of the dipterocarp family (Dipterocarpaceae).

Major Ecological regions

The World Wildlife Fund (WWF) divides Indomalayan realm into three bioregions, which it defines as "geographic clusters of ecoregions that may span several habitat types, but have strong biogeographic affinities, particularly at taxonomic levels higher than the species level (genus, family)."

Indian Subcontinent

The Indian Subcontinent bioregion covers most of India, Bangladesh, Nepal, Bhutan, and Sri Lanka. The Hindu Kush, Karakoram, Himalaya, and Patkai ranges bound the bioregion on the northwest, north, and northeast; these ranges were formed by the collision of the northward-drifting Indian subcontinent with Asia beginning 45 million years ago. The Hindu Kush, Karakoram, and Himalaya are a major biogeographic boundary between the subtropical and tropical flora and fauna of the Indian subcontinent and the temperate-climate Palearctic realm.

Indochina

The Indochina bioregion includes most of mainland Southeast Asia, including Myanmar, Thailand, Laos, Vietnam, and Cambodia, as well as the subtropical forests of southern China.

Sunda Shelf and the Philippines

Malesia is a botanical province which straddles the boundary between Indomalaya and Australasia. It includes the Malay Peninsula and the western Indonesian islands (known as Sundaland), the Philippines, the eastern Indonesian islands, and New Guinea. While the Malesia has much in common botanically, the portions east and west of the Wallace Line differ greatly in land animal species; Sundaland shares its fauna with mainland Asia, while terrestrial fauna on the islands east of the Wallace line are derived at least in part from species of Australian origin, such as marsupial mammals and ratite birds.

History

The flora of Indomalaya blends elements from the ancient supercontinents of Laurasia and Gondwana. Gondwanian elements were first introduced by India, which detached from Gondwana approximately 90 MYA, carrying its Gondwana-derived flora and fauna northward, which included cichlid fish and the flowering plant families Crypteroniaceae and possibly Dipterocarpaceae. India collided with Asia 30-45 MYA, and exchanged species. Later, as Australia-New Guinea drifted north, the collision of the Australian and Asian plates pushed up the islands of Wallacea, which were separated from one another by narrow straits, allowing a botanic exchange between Indomalaya and Australasia. Asian rainforest flora, including the dipterocarps, island-hopped across Wallacea to New Guinea, and several Gondwanian plant families, including podocarps and araucarias, moved westward from Australia-New Guinea into western Malesia and Southeast Asia.

Flora and Fauna

Two orders of mammals, the colugos (Dermoptera) and treeshrews (Scandentia), are endemic to the realm, as are families Craseonycteridae (Kitti's Hog-nosed Bat), Diatomyidae, Platacanthomyidae, Tarsiidae (tarsiers) and Hylobatidae (gibbons). Large mammals characteristic of Indomalaya include the leopard, tigers, water buffalos, Asian Elephant, Indian Rhinoceros, Javan Rhinoceros, Malayan Tapir, orangutans, and gibbons.

Indomalaya has three endemic bird families, the Irenidae (leafbirds and fairy bluebirds), Megalaimidae and Rhabdornithidae (Philippine creepers). Also characteristic are pheasants, pittas, Old World babblers, and flowerpeckers.

More information is available under Indomalayan realm fauna.

References

- Gargani J., Rigollet C. (2007). "Mediterranean Sea level variations during the Messinian Salinity Crisis.". Geophysical Research Letters. 34 (L10405): L10405. Bibcode:2007GeoRL..3410405G. doi:10.1029/2007GL029885

- Burgess, N.; D'Amico Hales, J.; Underwood, E.; et al., eds. (2004). Terrestrial Ecoregions of Africa and Madagascar: A Conservation Assessment. World Wildlife Fund Ecoregion Assessments (2nd ed.). Washington D.C.: Island Press. ISBN 978-1559633642. Archived from the original (PDF) on 2016-11-01

- Fox, Douglas (August 20, 2014). "Lakes under the ice: Antarctica's secret garden". Nature. 512: 244–246. Bibcode:2014Natur.512..244F. PMID 25143097. doi:10.1038/512244a. Retrieved August 21, 2014

- I.P.Farias et al.,Total Evidence: Molecules, Morphology, and the Phylogenetics of Cichlid Fishes, Journal of Experimental Zoology (Mol Dev Evol) 288 :76–92 (2000)

- Mack, Eric (August 20, 2014). "Life Confirmed Under Antarctic Ice; Is Space Next?". Forbes. Retrieved August 21, 2014

- Govers, R. (2009). Choking the Mediterranean to dehydration: The Messinian salinity crisis Geology, 37 (2), 167-170

3

An Integrated Study of Biome

A biome is classified as an area where the animals and plants have common characteristics. The common characteristics are mainly influenced by the climatic conditions of the area. The topics discussed in the chapter are of great importance to broaden the existing knowledge on biome.

Biome

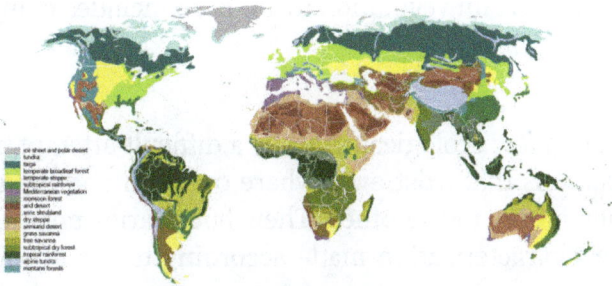

One way of mapping terrestrial biomes around the world

A biome is a community of plants and animals that have common characteristics for the environment they exist in, and can be found over a range of continents. Spanning continents, biomes are distinct biological communities that have formed in response to a shared physical climate. "Biome" is a broader term than "habitat"; any biome can comprise a variety of habitats.

While a biome can cover large areas, a microbiome is a mix of organisms that coexist in a defined space as well, but on a much smaller scale. For example, the human microbiome is the collection of bacteria, viruses, and other microorganisms that are present on a human.

A 'biota' is the total collection of organisms of a geographic region or a time period, from local geographic scales and instantaneous temporal scales all the way up to whole-planet and whole-timescale spatiotemporal scales. The biotas of the Earth make up the biosphere.

History of the Concept

The term was suggested in 1916 by Clements, originally as a synonym for biotic community of Möbius (1877). Later, it gained its current definition, based on earlier concepts of phytophysiognomy, formation and vegetation (used in opposition to flora), with the inclusion of the animal element and the exclusion of the taxonomic element of species composition. In 1935, Tansley added the climatic and soil aspects to the idea, calling it ecosystem. The International Biological Program (1964–74) projects popularized the concept of biome.

However, in some contexts, the term biome is used in a different manner. In German literature, particularly in the Walter terminology, the term is used similarly as biotope (a concrete geographical unit), while the biome definition used is used as an international, non-regional, terminology - irrespectively of the continent in which an area is present, it takes the same biome name - and corresponds to his "zonobiome", "orobiome" and "pedobiome" (biomes determinated by climate zone, altitude or soil).

In Brazilian literature, the term "biome" is sometimes used as synonym of "biogeographic province", an area based on species composition (the term "floristic province" being used when plant species are considered), or also as synonym of the "morphoclimatic and phytogeographical domain" of Ab'Sáber, a geographic space with subcontinental dimensions, with the predominance of similar geomorphologic and climatic characteristics, and of a certain vegetation form. Both includes many biomes in fact.

Classifications

To divide the world in a few ecological zones is a difficult attempt, notably because of the small-scale variations that exist everywhere on earth and because of the gradual changeover from one biome to the other. Their boundaries must therefore be drawn arbitrarily and their characterization made according to the average conditions that predominate in them.

A 1978 study on North American grasslands found a positive logistic correlation between evapotranspiration in mm/yr and above-ground net primary production in g/m²/yr. The general results from the study were that precipitation and water use led to above-ground primary production, while solar irradiation and temperature lead to below-ground primary production (roots), and temperature and water lead to cool and warm season growth habit. These findings help explain the categories used in Holdridge's bioclassification scheme, which were then later simplified by Whittaker. The number of classification schemes and the variety of determinants used in those schemes, however, should be taken as strong indicators that biomes do not fit perfectly into the classification schemes created.

Holdridge (1947, 1964) Life Zones

Holdridge classified climates based on the biological effects of temperature and rainfall on vegetation under the assumption that these two abiotic factors are the largest determinants of the types of vegetation found in a habitat. Holdridge uses the four axes to define 30 so-called "humidity provinces", which are clearly visible in his diagram. While this scheme largely ignores soil and sun exposure, Holdridge acknowledged that these were important.

Allee (1949) Biome-types

The principal biome-types by Allee (1949):

- Tundra
- Taiga
- Deciduous forest
- Grasslands
- Desert
- High plateaus
- Tropical forest
- Minor terrestrial biomes

Kendeigh (1961) Biomes

The principal biomes of the world by Kendeigh (1961):

A. Terrestrial

- o Temperate deciduous forest
- o Coniferous forest
- o Woodland
- o Chaparral
- o Tundra
- o Grassland
- o Desert
- o Tropical savanna
- o Tropical forest

B. Marine

- o Oceanic plankton and nekton
- o Balanoid-gastropod-thallophyte
- o Pelecypod-annelid
- o Coral reef

Whittaker (1962, 1970, 1975) Biome-types

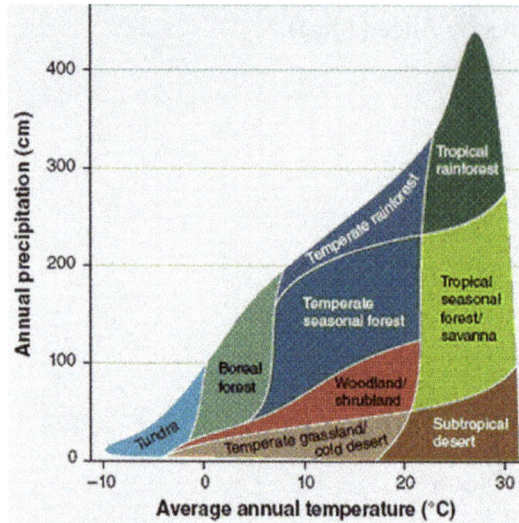

The distribution of vegetation types as a function of mean annual temperature and precipitation.

Whittaker classified biomes using two abiotic factors: precipitation and temperature. His scheme can be seen as a simplification of Holdridge's; more readily accessible, but missing Holdridge's greater specificity.

Whittaker based his approach on theoretical assertions and empirical sampling. He was in a unique position to make such a holistic assertion because he had previously compiled a review of biome classifications.

Key Definitions for Understanding Whittaker's Scheme

- Physiognomy: the apparent characteristics, outward features, or appearance of ecological communities or species.

- Biome: a grouping of terrestrial ecosystems on a given continent that are similar in vegetation structure, physiognomy, features of the environment and characteristics of their animal communities.

- Formation: a major kind of community of plants on a given continent.

- Biome-type: grouping of convergent biomes or formations of different continents, defined by physiognomy.

- Formation-type: a grouping of convergent formations.

Whittaker's distinction between biome and formation can be simplified: formation is used when applied to plant communities only, while biome is used when concerned with both plants and animals. Whittaker's convention of biome-type or formation-type is simply a broader method to categorize similar communities.

Whittaker's Parameters for Classifying Biome-types

Whittaker, seeing the need for a simpler way to express the relationship of community structure to the environment, used what he called "gradient analysis" of ecocline patterns to relate communities to climate on a worldwide scale. Whittaker considered four main ecoclines in the terrestrial realm.

1. Intertidal levels: The wetness gradient of areas that are exposed to alternating water and dryness with intensities that vary by location from high to low tide

2. Climatic moisture gradient

3. Temperature gradient by altitude

4. Temperature gradient by latitude

Along these gradients, Whittaker noted several trends that allowed him to qualitatively establish biome-types:

-   The gradient runs from favorable to extreme, with corresponding changes in productivity.

-   Changes in physiognomic complexity vary with how favorable of an environment exists (decreasing community structure and reduction of stratal differentiation as the environment becomes less favorable).

-   Trends in diversity of structure follow trends in species diversity; alpha and beta species diversities decrease from favorable to extreme environments.

-   Each growth-form (i.e. grasses, shrubs, etc.) has its characteristic place of maximum importance along the ecoclines.

-   The same growth forms may be dominant in similar environments in widely different parts of the world.

Whittaker summed the effects of gradients (3) and (4) to get an overall temperature gradient, and combined this with gradient (2), the moisture gradient, to express the above conclusions in what is known as the Whittaker classification scheme. The scheme graphs average annual precipitation (x-axis) versus average annual temperature (y-axis) to classify biome-types.

Biome-types

1. Tropical rainforest

2. Tropical seasonal rainforest

    o   deciduous

    o   semideciduous

3. Temperate giant rainforest

4. Montane rainforest

5. Temperate deciduous forest

6. Temperate evergreen forest

    o   needleleaf

    o   sclerophyll

7. Subarctic-subalpin needle-leaved forests (taiga)

8. Elfin woodland

9. Thorn forests and woodlands

10. Thorn scrub

11. Temperate woodland

12. Temperate shrublands

    o   deciduous

    o   heath

    o   sclerophyll

    o   subalpine-needleleaf

    o   subalpine-broadleaf

13. Savanna

14. Temperate grassland

15. Alpine grasslands

16. Tundra

17. Tropical desert

18. Warm-temperate desert

19. Cool temperate desert scrub

20. Arctic-alpine desert

21. Bog

22. Tropical fresh-water swamp forest

23. Temperate fresh-water swamp forest

24. Mangrove swamp

25. Salt marsh

26. Wetland

Walter (1976, 2002) Zonobiomes

The eponymously-named Heinrich Walter classification scheme considers the seasonality of temperature and precipitation. The system, also assessing precipitation and temperature, finds nine major biome types, with the important climate traits and vegetation types. The boundaries of each biome correlate to the conditions of moisture and cold stress that are strong determinants of plant form, and therefore the vegetation that defines the region. Extreme conditions, such as flooding in a swamp, can create different kinds of communities within the same biome.

Zonobiome	Zonal soil type	Zonal vegetation type
ZB I. Equatorial, always moist, little temperature seasonality	Equatorial brown clays	Evergreen tropical rainforest
ZB II. Tropical, summer rainy season and cooler "winter" dry season	Red clays or red earths	Tropical seasonal forest, seasonal dry forest, scrub, or savanna
ZB III. Subtropical, highly seasonal, arid climate	Serosemes, sierozemes	Desert vegetation with considerable exposed surface
ZB IV. Mediterranean, winter rainy season and summer drought	Mediterranean brown earths	Sclerophyllous (drought-adapted), frost-sensitive shrublands and woodlands
ZB V. Warm temperate, occasional frost, often with summer rainfall maximum	Yellow or red forest soils, slightly podsolic soils	Temperate evergreen forest, somewhat frost-sensitive
ZB VI. Nemoral, moderate climate with winter freezing	Forest brown earths and grey forest soils	Frost-resistant, deciduous, temperate forest
ZB VII. Continental, arid, with warm or hot summers and cold winters	Chernozems to serozems	Grasslands and temperate deserts
ZB VIII. Boreal, cold temperate with cool summers and long winters	Podsols	Evergreen, frost-hardy, needle-leaved forest (taiga)
ZB IX. Polar, short, cool summers and long, cold winters	Tundra humus soils with solifluction (permafrost soils)	Low, evergreen vegetation, without trees, growing over permanently frozen soils

Schultz (1988) Ecozones

Schultz (1988) defined nine ecozones (note that his concept of ecozone is more similar to the concept of biome used than to the concept of ecozone of BBC):

1. polar/subpolar zone

2. boreal zone

3. humid mid-latitudes

4. arid mid-latitudes

5. tropical/subtropical arid lands

6. Mediterranean-type subtropics

7. seasonal tropics

8. humid subtropics

9. humid tropics

Bailey (1989) Ecoregions

Robert G. Bailey nearly developed a biogeographical classification system of ecoregions for the United States in a map published in 1976. He subsequently expanded the system to include the rest of North America in 1981, and the world in 1989. The Bailey system, based on climate, is divided into seven domains (polar, humid temperate, dry, humid, and humid tropical), with further divisions based on other climate characteristics (subarctic, warm temperate, hot temperate, and subtropical; marine and continental; lowland and mountain).

- 100 Polar Domain

  o 120 Tundra Division (Köppen: Ft)

  o M120 Tundra Division – Mountain Provinces

  o 130 Subarctic Division (Köppen: E)

  o M130 Subarctic Division – Mountain Provinces

- 200 Humid Temperate Domain

  o 210 Warm Continental Division (Köppen: portion of Dcb)

  o M210 Warm Continental Division – Mountain Provinces

  o 220 Hot Continental Division (Köppen: portion of Dca)

  o M220 Hot Continental Division – Mountain Provinces

  o 230 Subtropical Division (Köppen: portion of Cf)

  o M230 Subtropical Division – Mountain Provinces

- o 240 Marine Division (Köppen: Do)

- o M240 Marine Division – Mountain Provinces

- o 250 Prairie Division (Köppen: arid portions of Cf, Dca, Dcb)

- o 260 Mediterranean Division (Köppen: Cs)

- o M260 Mediterranean Division – Mountain Provinces

- 300 Dry Domain

  - o 310 Tropical/Subtropical Steppe Division

  - o M310 Tropical/Subtropical Steppe Division – Mountain Provinces

  - o 320 Tropical/Subtropical Desert Division

  - o 330 Temperate Steppe Division

  - o 340 Temperate Desert Division

- 400 Humid Tropical Domain

  - o 410 Savanna Division

  - o 420 Rainforest Division

Olson & Dinerstein (1998) Biomes for WWF / Global 200

A team of biologists convened by the World Wildlife Fund (WWF) developed a scheme that divided the world's land area into biogeographic realms (called "ecozones" in a BBC scheme), and these into ecoregions (Olson & Dinerstein, 1998, etc.). Each ecoregion is characterized by a main biome (also called major habitat type).

This classification is used to define the Global 200 list of ecoregions identified by the WWF as priorities for conservation.

For the terrestrial ecoregions, there is a specific EcoID, format XXnnNN (XX is the biogeographic realm, nn is the biome number, NN is the individual number).

Biomes (terrestrial)

1. Tropical and subtropical moist broadleaf forests (tropical and subtropical, humid)

2. Tropical and subtropical dry broadleaf forests (tropical and subtropical, semi-humid)

3. Tropical and subtropical coniferous forests (tropical and subtropical, semihumid)

4. Temperate broadleaf and mixed forests (temperate, humid)

5. Temperate coniferous forests (temperate, humid to semihumid)

6. Boreal forests/taiga (subarctic, humid)

7. Tropical and subtropical grasslands, savannas, and shrublands (tropical and subtropical, semiarid)

8. Temperate grasslands, savannas, and shrublands (temperate, semiarid)

9. Flooded grasslands and savannas (temperate to tropical, fresh or brackish water inundated)

10. Montane grasslands and shrublands (alpine or montane climate)

11. Tundra (Arctic)

12. Mediterranean forests, woodlands, and scrub or sclerophyll forests (temperate warm, semihumid to semiarid with winter rainfall)

13. Deserts and xeric shrublands (temperate to tropical, arid)

14. Mangrove (subtropical and tropical, salt water inundated)

Biomes (freshwater)

According to the WWF, the following are classified as freshwater biomes:

- Large lakes

- Large river deltas

- Polar freshwaters

- Montane freshwaters

- Temperate coastal rivers

- Temperate floodplain rivers and wetlands

- Temperate upland rivers

- Tropical and subtropical coastal rivers

- Tropical and subtropical floodplain rivers and wetlands

- Tropical and subtropical upland rivers

- Xeric freshwaters and endorheic basins

- Oceanic islands
- Streams and rivers

Biomes (marine)

Biomes of the coastal and continental shelf areas (neritic zone):

- Polar
- Temperate shelves and sea
- Temperate upwelling
- Tropical upwelling
- Tropical coral

Anthropogenic Biomes

Humans have altered global patterns of biodiversity and ecosystem processes. As a result, vegetation forms predicted by conventional biome systems can no longer be observed across much of Earth's land surface as they have been replaced by crop and rangelands or cities. Anthropogenic biomes provide an alternative view of the terrestrial biosphere based on global patterns of sustained direct human interaction with ecosystems, including agriculture, human settlements, urbanization, forestry and other uses of land. Anthropogenic biomes offer a new way forward in ecology and conservation by recognizing the irreversible coupling of human and ecological systems at global scales and moving us toward an understanding of how best to live in and manage our biosphere and the anthropogenic biomes we live in.

Major anthropogenic biomes:

- Dense settlements
- Croplands
- Rangelands
- Forested
- Indoor

Microbial Biomes

Endolithic Biomes

The endolithic biome, consisting entirely of microscopic life in rock pores and cracks, kilometers beneath the surface, has only recently been discovered, and does not fit well into most classification schemes.

Dermal Biome

The dermal biome is the living ecosystem that animals (including humans) have evolved, that permits them to live symbiotically and in balance with the microbes on and in them (the microbiome). This ecosystem consists of skin, follicles, hair, sebaceous glands, sweat glands, arrector pili muscles, peptides, proteins, lipids and its associated microbiota. A healthy dermal biome has several functions: it resists infection of pathogens, protects against moisture loss and water damage, dynamically regulates body temperature and supports the healthy renewal of skin through the epidermal cell life cycle.

- Infection Resistance: Commensal microbiota assist the dermal biome resist infection for pathogenic bacteria by i] out-competing pathogens for resources, ii] training or stimulating the host's immune system to defeat the pathogen, or iii] expressing substances that are directly hostile to the pathogen.

- Water-barrier regulation: The dermal biome regulates the water barrier – preventing moisture from escaping (except when expressed as sweat) and preventing environmental water from permeating the skin. Because environmental water can have a chilling effect to mammals and warm-blooded animals when kept in close proximity to the epidermis, the dermal biome also produces hydrophobic lipids that repel water.

- Temperature regulation: The dermal biome is responsible for thermoregulation. To regulate excess heat, the dermal biome activates the sweat glands, allowing for evaporative cooling as sweat evaporates. To regulate cooler temperatures, the arrector pili muscles contract, causing hairs to "stand up" (goosebumps), and thereby trap an insulating blanket of air close to the skin.

- Skin renewal: a healthy biome supports the replacement of skin through the life cycle of epidermal cells as they proliferate in the basal layers of the epidermis until they die and are shed (desquamation).

Tropical and Subtropical Moist Broadleaf Forests

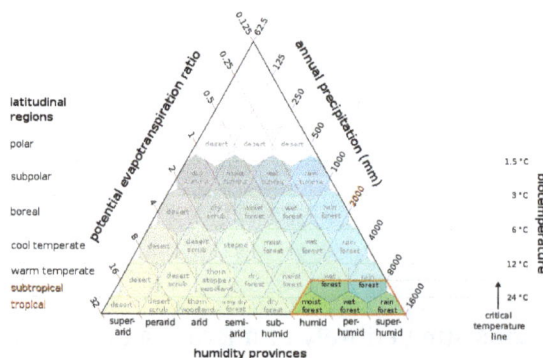

Tropical and subtropical moist forests (TSMF) as shown within the Holdridge Life Zones classification scheme, and includes moist forests, wet forests, and rainforests.

Tropical and subtropical moist broadleaf forests (TSMF), also known as tropical moist forests, are a tropical and subtropical forest biome (as defined under the *Global 200* scheme, promoted by the WWF); they may be referred to as jungle, especially when they are seasonal.

Definition

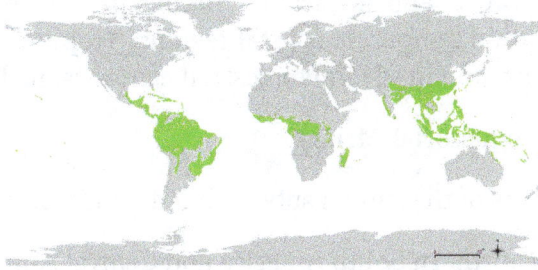

General distribution of tropical moist forests

The biome includes several types of forests:

- Lowland equatorial evergreen rain forests, commonly known as tropical rain-forests, are forests which receive high rainfall (tropical rainforest climate with more than 2000 mm, or 80 inches, annually) throughout the year. These forests occur in a belt around the equator, with the largest areas in the Amazon basin of South America, the Congo basin of central Africa, and parts of the Malay Archipelago.

Rainforest lining a river bank, Cameroon

- Tropical seasonal forest: also known as moist deciduous, monsoon or semi-ev-ergreen (mixed) seasonal forests have a monsoon or wet savannah climates (as in the Köppen climate classification): receiving high overall rainfall with a warm summer wet season and (often) a cooler winter dry season. Some trees in these forests drop some or all of their leaves during the winter dry season. These for-ests are found in parts of South America, in Central America and around the Caribbean, in coastal West Africa, parts of the Indian subcontinent, and across much of Indochina.

- Montane rain forests, are found in cooler-climate mountainous areas. Those with elevations high enough to regularly encounter low-level cloud cover are known as cloud forests.

- Flooded forests, including freshwater swamp forests and peat swamp forests.

In contrast to TSMF, tropical forest regions with lower levels of rainfall are home to tropical dry broadleaf forests and tropical coniferous forests. Temperate rain forests also occur in certain humid temperate coastal regions. Tropical and subtropical moist broadleaf forests are common in several terrestrial ecozones, including parts of:

- The Afrotropic (equatorial Africa)

- Indomalaya (parts of the Indian subcontinent and Southeast Asia)

- The Neotropic (northern South America and Central America)

- Australasia (eastern Indonesia, New Guinea, northern and eastern Australia)

- Oceania (the tropical islands of the Pacific Ocean)

About half of the world's tropical rainforests are in the South American countries of Brazil and Peru. Rainforests now cover less than 6% of Earth's land surface. Scientists estimate that more than half of all the world's plant and animal species live in tropical rain forests.

Tropical and Subtropical Dry Broadleaf Forests

Trinidad and Tobago dry forest on Chacachacare showing the dry-season deciduous
nature of the vegetation

The tropical and subtropical dry broadleaf forest biome, also known as tropical dry forest, vine thicket, and vine scrub is located at tropical and subtropical latitudes. Though these forests occur in climates that are warm year-round, and may receive several hundred centimeters of rain per year, they have long dry seasons which last several months and vary with geographic location. These seasonal droughts have great impact on all living things in the forest.

Subtropical semi-evergreen seasonal forest in Doi Inthanon National Park,
Northern Thailand, at the end of the dry season.

Deciduous trees are predominate in most of these forests, and during the drought a leafless period occurs, which varies with species type. Because trees lose moisture through their leaves, the shedding of leaves allows trees such as teak and mountain ebony to conserve water during dry periods. The newly bare trees open up the canopy layer, enabling sunlight to reach ground level and facilitate the growth of thick underbrush. Trees on moister sites and those with access to ground water tend to be evergreen. Infertile sites also tend to support evergreen trees. Three tropical dry broadleaf forest ecoregions, the East Deccan dry evergreen forests, the Sri Lanka dry-zone dry evergreen forests, and the Southeastern Indochina dry evergreen forests, are characterized by evergreen trees.

Though they have less biologic diversity than rainforests, tropical dry forests are home to a wide variety of wildlife including monkeys, deer, large cats, parrots, various rodents, and ground dwelling birds. Mammalian biomass tends to be higher in dry forests than in rain forests, especially in Asian and African dry forests. Many of these species display extraordinary adaptations to the difficult climate.

This biome is alternately known as the tropical bane forest biome or the tropical and subtropical deciduous forest biome. Locally some of these forests are also called monsoon forests, and they tend to merge into savannas.

Geographical Variation

As with seasonal tropical forests, dry forests tend to exist in the drier areas north and south of the tropical rainforest belt, south or north of the subtropical deserts, generally in two bands: one between 10° and 20°N latitude and the other between 10° and 20°S latitude. The most diverse dry forests in the world occur in southern Mexico and in the Bolivian lowlands. The dry forests of the Pacific Coast of northwestern South America support a wealth of unique species due to their dry climate. The Maputaland-Pondoland bushland and thickets along the east coast of South Africa are diverse and support many endemic species. The dry forests of central India and Indochina are notable for their diverse large vertebrate faunas. Madagascar dry deciduous forests and New

Caledonia dry forests are also highly distinctive (pronounced endemism and a large number of relictual taxa) for a wide range of taxa and at higher taxonomic levels. Trees use underground water during the dry seasons.

Afrotropic

- Madagascar dry deciduous forests

  o AT0202 Madagascar dry deciduous forests

Australasia

- Nusa Tenggara Dry Forests (Indonesia)

  o AA0201 Lesser Sundas deciduous forests

  o AA0203 Sumba deciduous forests

  o AA0204 Timor and Wetar deciduous forests

- New Caledonia dry forests

  o AA0202 New Caledonia dry forests

Caribbean

- U.S. Virgin Islands Dry Forest

Indomalaya

- Indochina dry forests

  o IM0202 Central Indochina dry forests

- Chhota - Nagpur dry forests

  o IM0203 Chota Nagpur dry deciduous forests

Neartic

- Northwestern Mexico

  o NA0201 Sonoran-Sinaloan ecoregion

Neotropic

- Mexican dry forests

  o NT0201 Apure-Villavicencio dry forests

  o NT0204 Bajio dry forests

- o NT0205 Balsas dry forests
- o NT0227 Sierra de la Laguna dry forests
- Tumbesian - Andean valleys dry forests (Colombia, Ecuador, Peru)
- o NT0214 Ecuadorian dry forests
- o NT0221 Magdalena Valley dry forests
- o NT0223 Marañón dry forests
- o NT0232 Tumbes-Piura dry forests
- Chiquitano dry forests
- o NT0212 Chiquitano dry forests
- Atlantic dry forests
- o NT0202 Atlantic dry forests

Oceania

- Hawaii, United States dry forests
- o OC0202 Hawaiian tropical dry forests

Biodiversity Patterns and Requirements

Species tend to have wider ranges than moist forest species, although in some regions many species do display highly restricted ranges; most dry forest species are restricted to tropical dry forests, particularly in plants; beta diversity and alpha diversity high but typically lower than adjacent moist forests.

Effective conservation of dry broadleaf forests requires the preservation of large and continuous areas of forest. Large natural areas are required to maintain larger predators and other vertebrates, and to buffer sensitive species from hunting pressure. The persistence of riparian forests and water sources is critical for many dry forest species. Large swathes of intact forest are required to allow species to recover from occasional large events, like forest fires.

Dry forests are highly sensitive to excessive burning and deforestation; overgrazing and exotic species can also quickly alter natural communities; restoration is possible but challenging, particularly if degradation has been intense and persistent.

Temperate Broadleaf and Mixed Forest

Temperate broadleaf and mixed forest is a temperate climate terrestrial biome, with broadleaf tree ecoregions, and with conifer and broadleaf tree mixed coniferous forest ecoregions.

Temperate broadleaf and mixed forest in Diqing Prefecture, Yunnan, southwest China.

The term 'Temperate broadleaf and mixed forest' is used by the World Wildlife Fund (WWF) in global biogeography as one of the biome designations under which to organize ecoregions.

Ecology

The typical structure of these forests includes four layers. The uppermost layer is the canopy composed of tall mature trees ranging from 30 to 61 m (100 to 200 ft) high. Below the canopy is the three-layered, shade-tolerant understory that is roughly 9 to 15 m (30 to 50 ft) shorter than the canopy. The top layer of the understory is the sub-canopy composed of smaller mature trees, saplings, and suppressed juvenile canopy layer trees awaiting an opening in the canopy. Below the sub-canopy is the shrub layer, composed of low growing woody plants. Typically the lowest growing (and most diverse) layer is the ground cover or herbaceous layer.

Trees

In the Northern hemisphere, characteristic dominant broadleaf trees in this biome include oaks (*Quercus* spp.), beeches (*Fagus* spp.), maples (*Acer* spp.), or birches (*Betula* spp.). The term "mixed forest" comes from the inclusion of coniferous trees as a canopy component of some of these forests. Typical coniferous trees include: Pines (*Pinus* spp.), firs (*Abies* spp.), and spruces (*Picea* spp.). In some areas of this biome the conifers may be a more important canopy species than the broadleaf species. In the Southern hemisphere, endemic genera such as *Nothofagus* and *Eucalyptus* occupy this biome.

Climate

Temperate broadleaf and mixed forests occur in areas with distinct warm and cool season, which give it a moderate annual average temperature — 3 to 15.6 °C (37 to 60 °F). These forests occur in relatively warm and rainy climates, sometimes also with a distinct dry season. A dry season occurs in the winter in East Asia and in summer on the wet fringe of the Mediterranean climate zones. Other areas, as in the central and upper eastern United States and southeastern Canada, have a fairly even distribution of rainfall; annual rainfall is typically over 600 mm (24 in) and often over 1,500 mm (59 in). Temperatures are typically moderate except in parts of Asia such as Ussuriland where temperate forests can occur despite very harsh conditions with very cold winters.

Montane Grasslands and Shrublands

Montane grasslands and shrublands is a biome defined by the World Wildlife Fund. The biome includes high altitude grasslands and shrublands around the world. The term "montane" in the name of the biome refers to "high altitude", rather than the ecological term which denotes the region below treeline.

Montane grasslands and shrublands located above the tree line are commonly known as alpine tundra, which occurs in mountain regions around the world. Below the tree line are subalpine and montane grasslands and shrublands. Stunted subalpine forests are known as krummholz, and occur just below the tree line, where harsh, windy conditions and poor soils create dwarfed and twisted forests of slow-growing trees.

Montane grasslands and shrublands, particularly in subtropical and tropical regions, often evolved as virtual islands, separated from other montane regions by warmer, lower elevation regions, and are frequently home to many distinctive and endemic plants which evolved in response to the cool, wet climate and abundant tropical sunlight. Characteristic plants of these habitats display adaptations such as rosette structures, waxy surfaces, and hairy leaves. A unique feature of many wet tropical montane regions is the presence of giant rosette plants from a variety of plant families, such as *Lobelia* (Afrotropic), *Puya* (Neotropic), *Cyathea* (New Guinea), and *Argyroxiphium* (Hawaii).

The most extensive montane grasslands and shrublands occur in the Neotropic Páramo of the Andes Mountains. This biome also occurs in the mountains of east and central Africa, Mount Kinabalu of Borneo, highest elevations of the Western Ghats in South India and the Central Highlands of New Guinea.

Where conditions are drier, one finds montane grasslands, savannas, and woodlands, like the Ethiopian Highlands, and montane steppes, like the steppes of the Tibetan Plateau.

Anthropogenic Biome

Anthropogenic biomes, also known as anthromes or human biomes, describe the terrestrial biosphere in its contemporary, human-altered form using global ecosystem units defined by global patterns of sustained direct human interaction with ecosystems. Anthromes were first named and mapped by Erle Ellis and Navin Ramankutty in their 2008 paper, "Putting People in the Map: Anthropogenic Biomes of the World". Anthrome maps now appear in numerous textbooks and in the National Geographic World Atlas

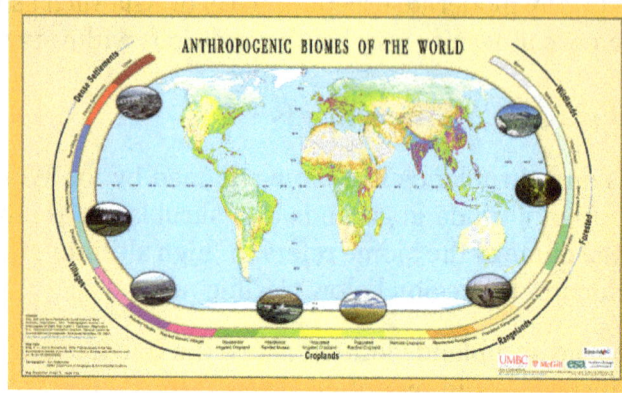

Anthropogenic biomes

Anthropogenic Transformation of the Biosphere

For more than a century, the biosphere has been described in terms of global ecosystem units called biomes, which are vegetation types like tropical rainforests and grasslands that are identified in relation to global climate patterns. Considering that human populations and their use of land have fundamentally altered global patterns of ecosystem form, process, and biodiversity, anthropogenic biomes provide a framework for integrating human systems with the biosphere in the Anthropocene.

Agriculture (1700-present)

Humans have been altering ecosystems largely since agriculture first developed, and as the human population has grown and become more technologically advanced over time, the land use for agricultural purposes has increased significantly. The anthropogenic biome in the 1700s, before the industrial revolution, was made up of mostly wild, untouched land, with no human settlement disturbing the natural state. In this time period, most of the Earth's ice-free land consisted of wildlands and natural anthromes, and it wasn't until after the industrial revolution in the 19th century that land use for agriculture and human settlements started to increase. With technology advancing and manufacturing processes becoming more efficient, the human population was beginning to thrive, and was subsequently requiring and using more natural resources. By the year 2000, over half of the Earth's ice free land was transformed into rangelands, croplands, villages and dense settlements, which left less than half of the Earth's land untouched. Anthropogenic changes between 1700 and 1800 were far smaller than those of the following centuries, and as such the rate of change has increased over time. As a result, the 20th century had the fastest rate of anthropogenic ecosystem transformation of the past 300 years.

Land Distribution

As the human population steadily increased in numbers throughout history, the use of natural resources and land began to increase, and the distribution of land used for various agricultural and settlement purposes began to change. The use of land around the world was transformed from its natural state to land used for agriculture, settlements and pastures to sustain the population and its growing needs. The distribution of land among anthromes underwent a shift away from natural anthromes and wildlands towards human-altered anthromes we are familiar with today. Now, the most populated anthromes (dense settlements and villages) account for only a small fraction of the global ice-free land. From the year 1700-2000, lands used for agriculture and urban settlements increased significantly, however the area occupied by rangelands increased even more rapidly, so that it became the dominant anthrome in the 20th century. As a result, the biggest global land-use change as a result of the industrial revolution, was the expansion of pastures.

Human Population

Following the industrial revolution, the human population experienced a rapid increase. The human population density in certain anthromes began to change, shifting away from rural environments to urban settlements, where the population density was much higher. These changes in population density between areas shifted global patterns of anthrome emergence, and also had wide-spread effects on various ecosystems. Half of the Earth's population now lives in cities, and most people reside in urban anthromes, with some populations dwelling in smaller cities and towns. Currently, human populations are expected to grow until at least midcentury, and the transformation of the Earth's anthromes are expected to follow this growth.

Current State of the Anthropogenic Biosphere

The present state of the terrestrial biosphere is predominantly anthropogenic. More than half of the terrestrial biosphere remains unused directly for agriculture or urban settlements, and of these unused lands still remaining, less than half are wildlands. Most of Earth's unused lands are now within the agricultural and settled landscapes of semi-natural, rangeland, cropland and village anthromes.

Major Anthromes

Anthromes include dense settlements (urban and mixed settlements), villages, croplands, rangelands and semi-natural lands and have been mapped globally using two different classification systems, viewable on Google Maps and Google Earth. There are currently 18 anthropogenic biomes.

Dense Settlements

Dense Settlements are the second most densely populated regions in the world. They are defined as areas with a high population density, though the density can be variable. The Population density, however, never falls below 100 persons/km, even in the non-urban parts of the dense settlements, and it has been suggested that these areas consist of both the edges of major cities in underdeveloped nations, and the long standing small towns throughout western Europe and Asia. Most often we think of dense settlements as cities, but dense settlements can also be suburbs, towns and rural settlements with high but fragmented populations.

Croplands

Croplands are another major anthrome throughout the world. Croplands include most of the cultivated lands of the world, and also about a quarter of global tree cover. Croplands which are locally irrigated have the highest human population density, likely due to the fact that it provides crops with a constant supply on water. This makes harvest time and crop survival more predictable. Croplands that are sustained mainly from the local rainfall are the most extensive of the populated anthromes, with annual precipitation near 1000 mm in certain areas of the globe. In these areas, there is sufficient water supplied by the climate to support all aspects of life without hardly any irrigation. However, in dryer areas, this method of agriculture would not be as productive.

Rangelands

Rangelands are a very broad anthropogenic biome group that has been described according to three levels of population density: residential, populated and remote. The Residential rangeland anthrome has two key features: its population density is never below 10 persons/km and a substantial portion of its area is used for pasture. Pastures in rangelands are the most dominant land cover. Bare earth is significant in this anthrome, covering nearly one third of the land for every one square kilometer. Rangeland anthromes are less altered than croplands, but their alteration tends to increase with population. Domesticated grazing livestock are typically adapted to grasslands and savannas, so the alteration of these biomes tends to be less noticeable.

Forests

Forested anthromes are dominated by tree cover, and they have high precipitation and minimal human populations, where the population density is usually less than 3 persons/km^2. Most populated forests act as carbon sinks because of the lack of human activity. Without harmful emissions being released in the forests due to human activity, the vegetation is able to utilize carbon dioxide in the atmosphere, and act as a sink. Remote forests are a little different than populated forests because the majority of the vegetation in these forests have been clear-cut for human consumption. Forests are generally cleared to sustain substantial populations of domestic livestock, and to utilize the lumber.

Indoor

Very few biologists have studied the evolutionary processes at work in indoor environments. Estimates of the extent of residential and commercial buildings range between 1.3% and 6% of global ice-free land area. This area is just as extensive as other small biomes such as flooded grass-lands and tropical coniferous forests. The indoor biome is rapidly expanding, while forest anthromes are shrinking. The indoor biome of Manhattan is almost three times as large, in terms of its floor space, as is the geographical area of the island itself, due to the buildings rising up instead of spreading out. Thousands of species live in the indoor biome, many of them preferentially or even obligatorily. The only action that humans take to alter the evolution of the indoor biome is with cleaning practices. The field of indoor biomes will continue to change as long as our culture will change.

Implications of an Anthropogenic Biosphere

Humans have fundamentally altered global patterns of biodiversity and ecosystem processes. It is no longer possible to explain or predict ecological patterns or processes across the Earth without considering the human role. Human societies began transforming terrestrial ecology more than 50000 years ago, and evolutionary evidence has been presented demonstrating that the ultimate causes of human transformation of the biosphere are social and cultural, not biological, chemical, or physical. Anthropogenic biomes offer a new way forward by acknowledging human influence on global ecosystems and moving us toward models and investigations of the terrestrial biosphere that integrate human and ecological systems.

Challenges Facing Biodiversity in the Anthropogenic Biosphere

Extinctions

Over the past century, anthrome extent and land use intensity increased rapidly together with growing human populations, leaving wildlands without human population or land use in less than one quarter of the terrestrial biosphere. This massive transformation of Earth's ecosystems for human use has occurred with enhanced rates of species extinctions. Humans are directly causing species extinctions, especially of megafauna, by reducing, fragmenting and transforming native habitats and by overexploiting individual species. Current rates of extinctions vary greatly by taxa, with mammals, reptiles and amphibians especially threatened; however there is growing evidence that viable populations of many, if not most native taxa, especially plants, may be sustainable within anthromes. With the exception of especially vulnerable taxa, the majority of native species may be capable of maintaining viable populations in anthromes.

Conservation

Anthromes present an alternative view of the terrestrial biosphere by characterizing the diversity of global ecological land cover patterns created and sustained by human population densities and land use while also incorporating their relationships with biotic communities. Biomes and ecoregions are limited in that they reduce human influences, and an increasing number of conservation biologists have argued that biodiversity conservation must be extended to habitats directly shaped by humans. Within anthromes, including densely populated anthromes, humans rarely use all available land. As a result, anthromes are generally mosaics of heavily used lands and less intensively used lands. Protected areas and biodiversity hotspots are not distributed equally across anthromes. Less populated anthromes contain a greater proportion of protected areas. While 23.4% of remote woodland anthrome is protected, only 2.3% of irrigated village anthrome is protected. There is increasing evidence that suggests that biodiversity conservation can be effective in both densely and sparsely settled anthromes. Land sharing and land sparing are increasingly seen as conservation strategies.

Pinophyta

The Pinophyta, also known as Coniferophyta or Coniferae, or commonly as conifers, are a division of vascular land plants containing a single class, Pinopsida. They are gymnosperms, cone-bearing seed plants. All extant conifers are perennial woody plants with secondary growth. The great majority are trees, though a few are shrubs. Examples include cedars, Douglas firs, cypresses, firs, junipers, kauri, larches, pines, hemlocks, redwoods, spruces, and yews. As of 1998, the division Pinophyta was estimated to contain eight families, 68 genera, and 629 living species.

Although the total number of species is relatively small, conifers are ecologically important. They are the dominant plants over large areas of land, most notably the taiga of the Northern Hemisphere, but also in similar cool climates in mountains further south. Boreal conifers have many wintertime adaptations. The narrow conical shape of northern conifers, and their downward-drooping limbs, help them shed snow. Many of them seasonally alter their biochemistry to make them more resistant to freezing. While tropical rainforests have more biodiversity and turnover, the immense conifer forests of the world represent the largest terrestrial carbon sink. Conifers are of great economic value for softwood lumber and paper production.

Evolution

The earliest conifers in the fossil record date to the late Carboniferous (Pennsylvanian) period (about 300 million years ago), possibly arising from *Cordaites*, a genus of seed-bearing Gondwanan plants with cone-like fertile structures. Pinophytes, Cycadophytes, and Ginkgophytes all developed at this time. An important adaptation of these gymnosperms was allowing plants to live without being so dependent on water. Other adaptations are pollen (so fertilization can occur without water) and the seed, which allows the embryo to be transported and developed elsewhere.

The narrow conical shape of northern conifers, and their downward-drooping limbs, help them shed snow.

Conifers appear to be one of the taxa that benefited from the Permian–Triassic extinction event, and were the dominant land plants of the Mesozoic. They were overtaken by the flowering plants, which first appeared in the Cretaceous, and became dominant in the Cenozoic era. They were the main food of herbivorous dinosaurs, and their resins and poisons would have given protection against herbivores. Reproductive features of modern conifers had evolved by the end of the Mesozoic era.

Taxonomy and Naming

Conifer is a Latin word, a compound of *conus* (cone) and *ferre* (to bear), meaning "the one that bears (a) cone(s)".

The division name Pinophyta conforms to the rules of the *International Code of Nomenclature for algae, fungi, and plants (ICN)*, which state (Article 16.1) that the names of higher taxa in plants (above the rank of family) are either formed from the name of an included family (usually the most common and/or representative), in this case Pinaceae (the pine family), or are descriptive. A descriptive name in widespread use for the conifers (at whatever rank is chosen) is Coniferae (Art 16 Ex 2).

According to the *ICN*, it is possible to use a name formed by replacing the termination -*aceae* in the name of an included family, in this case preferably Pinaceae, by the appropriate termination, in the case of this division *ophyta*. Alternatively, "descriptive botanical names" may also be used at any rank above family. Both are allowed.

This means that if conifers are considered a division, they may be called Pinophyta or

Coniferae. As a class they may be called Pinopsida or Coniferae. As an order they may be called Pinales or Coniferae or Coniferales.

Conifers are the largest and economically most important component group of the gymnosperms, but nevertheless they comprise only one of the four groups. The division Pinophyta consists of just one class, Pinopsida, which includes both living and fossil taxa. Subdivision of the living conifers into two or more orders has been proposed from time to time. The most commonly seen in the past was a split into two orders, Taxales (Taxaceae only) and Pinales (the rest), but recent research into DNA sequences suggests that this interpretation leaves the Pinales without Taxales as paraphyletic, and the latter order is no longer considered distinct. A more accurate subdivision would be to split the class into three orders, Pinales containing only Pinaceae, Araucariales containing Araucariaceae and Podocarpaceae, and Cupressales containing the remaining families (including Taxaceae), but there has not been any significant support for such a split, with the majority of opinion preferring retention of all the families within a single order Pinales, despite their antiquity and diverse morphology.

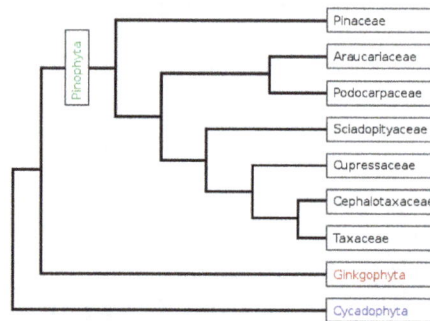

Phylogeny of the Pinophyta based on cladistic analysis of molecular data.

The conifers are now accepted as comprising seven families, with a total of 65–70 genera and 600–630 species (696 accepted names). The seven most distinct families are linked in the box above right and phylogenetic diagram left. In other interpretations, the Cephalotaxaceae may be better included within the Taxaceae, and some authors additionally recognize Phyllocladaceae as distinct from Podocarpaceae (in which it is included here). The family Taxodiaceae is here included in family Cupressaceae, but was widely recognized in the past and can still be found in many field guides. A new classification and linear sequence based on molecular data can be found in an article by Christenhusz et al.

The conifers are an ancient group, with a fossil record extending back about 300 million years to the Paleozoic in the late Carboniferous period; even many of the modern genera are recognizable from fossils 60–120 million years old. Other classes and orders, now long extinct, also occur as fossils, particularly from the late Paleozoic and Mesozoic eras. Fossil conifers included many diverse forms, the most dramatically distinct from modern conifers being some herbaceous conifers with no woody stems. Major fossil orders of conifers or conifer-like plants include the Cordaitales, Vojnovskyales, Voltziales and perhaps also the Czekanowskiales (possibly more closely related to the Ginkgophyta).

Morphology

All living conifers are woody plants, and most are trees, the majority having monopodial growth form (a single, straight trunk with side branches) with strong apical dominance. Many conifers have distinctly scented resin, secreted to protect the tree against insect infestation and fungal infection of wounds. Fossilized resin hardens into amber. The size of mature conifers varies from less than one meter, to over 100 meters. The world's tallest, thickest, largest, and oldest living trees are all conifers. The tallest is a Coast Redwood (*Sequoia sempervirens*), with a height of 115.55 meters (although one Victorian mountain ash, *Eucalyptus regnans*, allegedly grew to a height of 140 meters, although the exact dimensions were not confirmed). The thickest, or tree with the greatest trunk diameter, is a Montezuma Cypress (*Taxodium mucronatum*), 11.42 meters in diameter. The largest tree by three-dimensional volume is a Giant Sequoia (*Sequoiadendron giganteum*), with a volume 1486.9 cubic meters. The smallest is the pygmy pine (*Lepidothamnus laxifolius*) of New Zealand, which is seldom taller than 30 cm when mature. The oldest is a Great Basin Bristlecone Pine (*Pinus longaeva*), 4,700 years old.

Foliage

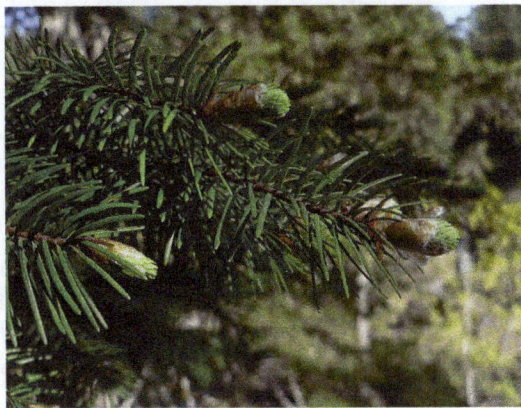

Pinaceae: needle-like leaves and vegetative buds of Coast Douglas fir
(*Pseudotsuga menziesii* var. *menziesii*)

Since most conifers are evergreens, the leaves of many conifers are long, thin and have a needle-like appearance, but others, including most of the Cupressaceae and some of the Podocarpaceae, have flat, triangular scale-like leaves. Some, notably *Agathis* in Araucariaceae and *Nageia* in Podocarpaceae, have broad, flat strap-shaped leaves. Others such as *Araucaria columnaris* have leaves that are awl-shaped. In the majority of conifers, the leaves are arranged spirally, exceptions being most of Cupressaceae and one genus in Podocarpaceae, where they are arranged in decussate opposite pairs or whorls of 3 (-4).

Araucariaceae: Awl-like leaves of Cook Pine (*Araucaria columnaris*)

In *Abies grandis* (*grand fir*), and many other species with spirally arranged leaves, leaf bases are twisted to flatten their arrangement and maximize light capture.

In many species with spirally arranged leaves, such as *Abies grandis* (pictured), the leaf bases are twisted to present the leaves in a very flat plane for maximum light capture. Leaf size varies from 2 mm in many scale-leaved species, up to 400 mm long in the needles of some pines (e.g. Apache Pine, *Pinus engelmannii*). The stomata are in lines or patches on the leaves, and can be closed when it is very dry or cold. The leaves are often dark green in colour, which may help absorb a maximum of energy from weak sunshine at high latitudes or under forest canopy shade. Conifers from hotter areas with high sunlight levels (e.g. Turkish Pine *Pinus brutia*) often have yellower-green leaves, while others (e.g. blue spruce, *Picea pungens*) have a very strong glaucous wax bloom to reflect ultraviolet light. In the great majority of genera the leaves are evergreen, usually remaining on the plant for several (2-40) years before falling, but five genera (*Larix*, *Pseudolarix*, *Glyptostrobus*, *Metasequoia* and *Taxodium*) are deciduous, shedding the leaves in autumn and leafless through the winter. The seedlings of many conifers, including most of the Cupressaceae, and *Pinus* in Pinaceae, have a distinct juvenile foliage period where the leaves are different, often markedly so, from the typical adult leaves.

Cupressaceae: scale leaves of Lawson's Cypress (*Chamaecyparis lawsoniana*); scale in mm

Tree Ring Structure

The internal structure of conifer

Tree rings are records of the influence of environmental conditions, their anatomical characteristics record growth rate changes produced by these changing conditions. The microscopic structure of conifer wood consists of two types of cells: parenchyma, which have an oval or polyhedral shape with approximately identical dimensions in three directions, and strongly elongated tracheids. Tracheids make up more than 90% of timber volume. The tracheids of earlywood formed at the beginning of a growing season have large radial sizes and smaller, thinner cell walls. Then, the first tracheids of the transition zone are formed, where the radial size of cells and thickness of their cell walls changes considerably. Finally, the latewood tracheids are formed, with small radial sizes and greater cell wall thickness. This is the basic pattern of the internal cel structure of conifer tree rings.

Reproduction

Most conifers are monoecious, but some are subdioecious or dioecious; all are wind-pollinated. Conifer seeds develop inside a protective cone called a strobilus. The cones take from four months to three years to reach maturity, and vary in size from 2 mm to 600 mm long.

Pinaceae: unopened female cones of subalpine fir (*Abies lasiocarpa*)

In Pinaceae, Araucariaceae, Sciadopityaceae and most Cupressaceae, the cones are woody, and when mature the scales usually spread open allowing the seeds to fall out and be dispersed by the wind. In some (e.g. firs and cedars), the cones disintegrate to release the seeds, and in others (e.g. the pines that produce pine nuts) the nut-like seeds are dispersed by birds (mainly nutcrackers, and jays), which break up the specially adapted softer cones. Ripe cones may remain on the plant for a varied amount of time before falling to the ground; in some fire-adapted pines, the seeds may be stored in closed cones for up to 60–80 years, being released only when a fire kills the parent tree.

Taxaceae: the fleshy aril that surrounds each seed in the European Yew (*Taxus baccata*) is a highly modified seed cone scale

In the families Podocarpaceae, Cephalotaxaceae, Taxaceae, and one Cupressaceae genus (*Juniperus*), the scales are soft, fleshy, sweet and brightly colored, and are eaten by fruit-eating birds, which then pass the seeds in their droppings. These fleshy scales are (except in *Juniperus*) known as arils. In some of these conifers (e.g. most Podocarpaceae), the cone consists of several fused scales, while in others (e.g. Taxaceae), the cone is reduced to just one seed scale or (e.g. Cephalotaxaceae) the several scales of a cone develop into individual arils, giving the appearance of a cluster of berries.

The male cones have structures called microsporangia that produce yellowish pollen through meiosis. Pollen is released and carried by the wind to female cones. Pollen grains from living pinophyte species produce pollen tubes, much like those of angiosperms. The gymnosperm male gametophytes (pollen grains) are carried by wind to a female cone and are drawn into a tiny opening on the ovule called the micropyle. It is within the ovule that pollen-germination occurs. From here, a pollen tube seeks out the female gametophyte and if successful, fertilization occurs. The resulting zygote develops into an embryo, which along with its surrounding integument, becomes a seed. Eventually the seed may fall to the ground and, if conditions permit, grow into a new plant.

In forestry, the terminology of flowering plants has commonly though inaccurately been applied to cone-bearing trees as well. The male cone and unfertilized female cone are called *male flower* and *female flower*, respectively. After fertilization, the female cone is termed *fruit*, which undergoes *ripening* (maturation).

It was found recently that the pollen of conifers transfers the mitochondrial organelles to the embryo, a sort of meiotic drive that perhaps explains why Pinus and other conifers are so productive, and perhaps also has bearing on (observed?) sex-ratio bias

Pinaceae: pollen cone of a Japanese Larch (*Larix kaempferi*)

Life Cycle

Conifers are heterosporous, generating two different types of spores: male microspores and female megaspores. These spores develop on separate male and female sporophylls on separate male and female cones. In the male cones, microspores are produced from microsporocytes by meiosis. The microspores develop into pollen grains, which are male gametophytes. Large amounts of pollen are released and carried by the wind. Some pollen grains will land on a female cone for pollination. The generative cell in the pollen grain divides into two haploid sperm cells by mitosis leading to the development of the pollen tube. At fertilization, one of the sperm cells unites its haploid nucleus with the haploid nucleus of an egg cell. The female cone develops two ovule, each of which contains haploid haploid megaspores. A megasporocyte is divided by meiosis in each ovule. Each winged pollen grain is a four celled male gametophyte Three of the four cells break down leaving only a single surviving cell which will develop into a female multicellular gametophyte. The female gametophytes grow to produce two or more archegonia, each of which contains an egg. Upon fertilization, the diploid egg will give rise to the embryo, and a seed is produced. The female cone then opens, releasing the seeds which grow to a young seedling.

1. To fertilize the ovum, the male cone releases pollen that is carried on the wind to the female cone. This is pollination. (Male and female cones usually occur on the same plant.)

2. The pollen fertilizes the female gamete (located in the female cone). Fertilization in some species does not occur until 15 months after pollination.

3. A fertilized female gamete (called a zygote) develops into an embryo.

4. A seed develops which contains the embryo. The seed also contains the integument cells surrounding the embryo. This is an evolutionary characteristic of the Spermatophyta.

5. Mature seed drops out of cone onto the ground.

6. Seed germinates and seedling grows into a mature plant.

7. When the plant is mature, it produces cones and the cycle continues.

Female Reproductive Cycles

Conifer reproduction is synchronous with seasonal changes in temperate zones. Reproductive development slows to a halt during each winter season, and then resumes each spring. The male strobilus development is completed in a single year. Conifers are classified by three reproductive cycles, namely; 1-, 2-, or 3- . The cycles refers to the completion of female strobilus development from initiation to seed maturation. All three types or reproductive cycles have a long gap in between pollination and fertilization.

One year reproductive cycle:The genera includes *Abies, Picea, Cedrus, Pseudotsuga, Tsuga, Keteleeria (Pinaceae)* and *Cupressus, Thuja, Cryptomeria, Cunninghamia* and *Sequoia (Cupressaceae)*. Female strobili are initiated in late summer or fall in a year, then they overwinter. Female strobili emerge followed by pollination in the following spring . Fertilization takes place in summer of the following year, only 3–4 months after pollination. Cones mature and seeds are then shed by the end of that same year. Pollination and fertilization occurs in a single growing season.

Two-year reproductive cycle:The genera includes *Widdringtonia, Sequoiadendron (Cupressaceae)* and most species of *Pinus*. Female strobilus initials are formed in late summer or fall then overwinter. It emerges and receives pollen in the first year spring and become conelets. The conelet goes through another winter rest and in the spring of the 2nd year. The Archegonia form in the conelet and fertilization of the archegonia occurs by early summer of the 2nd year, so the pollination-fertilization interval exceeds a year. After fertilization, the conelet is considered an immature cone. Maturation occurs by autumn of the 2nd year, at which time seeds are shed. In summary, the 1-year and the 2-year cycles differ mainly in the duration of the pollination- fertilization interval.

Three-year reproductive cycle: Three of the conifer species are pine species (*Pinus pinea*, *Pinus leiophylla*, *Pinus torreyana*) which have pollination and fertilization events separated by a 2-year interval. Female strobili initiated during late summer or autumn in a year, then overwinter until the following spring. Female strobili emerge then pollination occurs in spring of the 2nd year then the pollinated strobili become conelets in same year (i.e. the second year). The female gametophytes in the conelet develop so slowly that the megaspore does not go through free-nuclear divisions until autumn of the 3rd year. The conelet then overwinters again in the free-nuclear female gametophyte stage. Fertilization takes place by early summer of the 4th year and seeds mature in the cones by autumn of the 4th year.

Tree Development

The growth and form of a forest tree are the result of activity in the primary and secondary meristems, influenced by the distribution of photosynthate from its needles and the hormonal gradients controlled by the apical meristems (Fraser et al. 1964). External factors also influence growth and form.

Fraser recorded the development of a single white spruce tree from 1926 to 1961. Apical growth of the stem was slow from 1926 through 1936 when the tree was competing with herbs and shrubs and probably shaded by larger trees. Lateral branches began to show reduced growth and some were no longer in evidence on the 36-year-old tree. Apical growth totalling about 340 m, 370 m, 420 m, 450 m, 500 m, 600 m, and 600 m was made by the tree in the years 1955 through 1961, respectively. The total number of needles of all ages present on the 36-year-old tree in 1961 was 5.25 million weighing 14.25 kg. In 1961, needles as old as 13 years remained on the tree.The ash weight of needles increased progressively with age from about 4% in first-year needles in 1961 to about 8% in needles 10 years old. In discussing the data obtained from the one 11 m tall white spruce, Fraser et al. (1964) speculated that if the photosynthate used in making apical growth in 1961 was manufactured the previous year, then the 4 million needles that were produced up to 1960 manufactured food for about 600,000 mm of apical growth or 730 g dry weight, over 12 million mm³ of wood for the 1961 annual ring, plus 1 million new needles, in addition to new tissue in branches, bark, and roots in 1960. Added to this would be the photosynthate to produce energy to sustain respiration over this period, an mount estimated to be about 10% of the total annual photosynthate production of a young healthy tree. On this basis, one needle produced food for about

0.19 mg dry weight of apical growth, 3 mm³ wood, one-quarter of a new needle, plus an unknown amount of branch wood, bark and roots.

The order of priority of photosynthate distribution is probably: first to apical growth and new needle formation, then to buds for the next year's growth, with the cambium in the older parts of the branches receiving sustenance last. In the white spruce studied by Fraser et al. (1964), the needles constituted 17.5% of the over-day weight. Undoubtedly, the proportions change with time.

Seed Dispersal Mechanism

Wind and animals dispersals are two major mechanisms involved in the dispersal of conifer seeds. Wind bore seed dispersal involves two processes, namely; local neighborhood dispersal (LND) and long- distance dispersal (LDD). Long-distance dispersal distances ranges from 11.9–33.7 kilometres (7.4–20.9 mi) from the source. The bird family, Corvidae is the primary distributor of the conifer seeds. These birds are known to cache 32,000 pine seeds and transport the seeds as far as 12–22 kilometres (7.5–13.7 mi) from the source. The birds store the seeds in the soil at depths of 2–3 centimetres (0.79–1.18 in) under conditions which favor germination.

Invasive Species

A number of conifers originally introduced for forestry have become invasive species in parts of New Zealand, including Radiata pine (*Pinus radiata*), Lodgepole pine (*P. contorta*), Douglas fir (*Pseudotsuga mensiezii*) and European larch (*Larix decidua*). In parts of South Africa, *Pinus pinaster*, *P. patula* and *P. radiata* have been declared invasive species. These wilding conifers are a serious environmental issue causing problems for pastoral farming and for conservation.

Predators

At least 20 species of roundheaded borers of the family Cerambycidae feed on the wood of spruce, fir, and hemlock (Rose and Lindquist 1985). Borers rarely bore tunnels in living trees, although when populations are high, adult beetles feed on tender twig bark, and may damage young living trees. One of the most common and widely distributed borer species in North America is the whitespotted sawyer (*Monochamus scutellatus*). Adults are found in summer on newly fallen or recently felled trees chewing tiny slits in the bark in which they lay eggs. The eggs hatch in about 2 weeks, and the tiny larvae tunnel to the wood and score its surface with their feeding channels. With the onset of cooler weather, they bore into the wood making oval entrance holes and tunnel deeply. Feeding continues the following summer, when larvae occasionally return to the surface of the wood and extend the feeding channels generally in a U-shaped configuration. During this time, small piles of frass extruded by the larvae accumulate under logs. Early in the spring of the second year following egg-laying, the larvae, about 30 mm long, pupate in the tunnel enlargement just below the wood surface. The resulting adults chew their way out in early summer, leaving round exit holes, so completing the usual 2-year life cycle.

Cultivation

A cultivar of *Pinus sylvestris* with a narrow "fastigiate" growth habit

Conifers – notably *Abies* (fir), *Cedrus*, *Chamaecyparis lawsoniana* (Lawson's cypress), *Cupressus* (cypress), juniper, *Picea* (spruce), *Pinus* (pine), *Taxus* (yew), *Thuja* - have been the subject of selection for ornamental purposes (for more information see the silviculture page). Plants with unusual growth habits, sizes, and colours are propagated and planted in parks and gardens throughout the world.

Conditions for Growth

Conifers can absorb nitrogen in either the ammonium (NH_4^+) or nitrate (NO_3^-) form, but the forms are not physiologically equivalent. Form of nitrogen affected both the total amount and relative composition of the soluble nitrogen in white spruce tissues (Durzan and Steward 1967). Ammonium nitrogen was shown to foster arginine and amides and lead to a large increase of free guanidine compounds, whereas in leaves nourished by nitrate as the sole source of nitrogen guanidine compounds were less prominent. Durzan and Steward noted that their results, drawn from determinations made in late summer, did not rule out the occurrence of different interim responses at other times of year. Ammonium nitrogen produced significantly heavier (dry weight) seedlings with higher nitrogen content after 5 weeks (McFee and Stone 1968) than did the same amount of nitrate nitrogen. Swan (1960) found the same effect in 105-day-old white spruce.

The general short-term effect of nitrogen fertilization on coniferous seedlings is to stimulate shoot growth more so than root growth (Armson and Carman 1961). Over a longer period, root growth is also stimulated. Many nursery managers were long reluctant to apply nitrogenous fertilizers late in the growing season, for fear of increased danger of frost damage to succulent tissues. A presentation at the North American Forest Tree Nursery Soils Workshop at Syracuse in 1980 provided strong contrary evidence: Bob Eastman, President of the Western Maine Forest Nursery Co. stated that for 15 years he has been successful in avoiding winter "burn" to Norway spruce and white spruce in his nursery operation by fertilizing with 50-80 lb/ac (56–90 kg/ha) nitrogen in September, whereas previously winter burn had been experienced annually, often severely. Eastman also stated that the overwintering storage capacity of stock thus treated was much improved (Eastman 1980).

Of course, the concentrations of nutrient in plant tissues depend on many factors, including growing conditions. Interpretation of concentrations determined by analysis is easy only when a nutrient occurs in excessively low or occasionally excessively high concentration. Values are influenced by environmental factors and interactions among the 16 nutrient elements known to be essential to plants, 13 of which are obtained from the soil, including nitrogen, phosphorus, potassium, calcium, magnesium, and sulfur, all used in relatively large amounts (Buckman and Brady 1969). Nutrient concentrations in conifers also vary with season, age and kind of tissue sampled, and analytical technique. The ranges of concentrations occurring in well-grown plants provide a useful guide by which to assess the adequacy of particular nutrients, and the ratios among the major nutrients are helpful guides to nutritional imbalances.

Economic Importance

The softwood derived from conifers is of great economic value, providing about 45% of the world's annual lumber production. Other uses of the timber include the production of paper and plastic from chemically treated wood pulp. Some conifers also provide foods such as pine nuts and Juniper berries, used to flavor gin.

References

- Sims, Phillip L.; Singh, J.S. (July 1978). "The Structure and Function of Ten Western North American Grasslands: III. Net Primary Production, Turnover and Efficiencies of Energy Capture and Water Use". Journal of Ecology. British Ecological Society. 66 (2): 573–597. doi:10.2307/2259152

- Chapin III, F. Stuart; Matson, Pamela A; Vitousek, Peter M (2012). Principles of Terrestrial Ecosystem Ecology (Second ed.). Springer. ISBN 9781441995049

- Zimmer, Carl (March 19, 2015). "The Next Frontier: The Great Indoors". New York Times. Retrieved March 2015

- Smalls, Lola K., et al. "Quantitative model of cellulite: three-dimensional skin surface topography, biophysical characterization, and relationship to human perception." International Journal of Cosmetic Science 27.5 (2005): 295-295

- National Geographic Society (2014). National Geographic Atlas of the World (10th ed.). National Geographic. ISBN 9781426213540

- Lott, John N. A; Liu, Jessica C; Pennell, Kelly A; Lesage, Aude; West, M Marcia. "Iron-rich particles and globoids in embryos of seeds from phyla Coniferophyta, Cycadophyta, Gnetophyta, and Ginkgophyta: characteristics of early seed plants". 80 (9): 954–961. doi:10.1139/b02-083

- Lugo, A. E. (1999). "The Holdridge life zones of the conterminous United States in relation to ecosystem mapping" (PDF). Journal of Biogeography. 26: 1025–1038. doi:10.1046/j.1365-2699.1999.00329.x. Retrieved 27 May 2015

- Cain, Michael; Bowman, William; Hacker, Sally (2014). Ecology (Third ed.). Massachusetts: Sinauer. p. 51. ISBN 9780878939084

- Ledig, F. Thomas; Porterfield, Richard L., 1982, Tree Improvement in Western Conifers: Economic Aspects, Journal of Forestry

4

Fundamental Concepts of Biogeography

The fundamental concepts of biogeography are allopatric speciation, evolution, biological dispersal extinction and endemism. Allopatric speciation is the process that occurs when the population of the same species becomes secluded with each other resulting in genetic variations. This chapter discusses the fundamental concepts of biogeography in a critical manner providing key analysis to the subject matter.

Allopatric Speciation

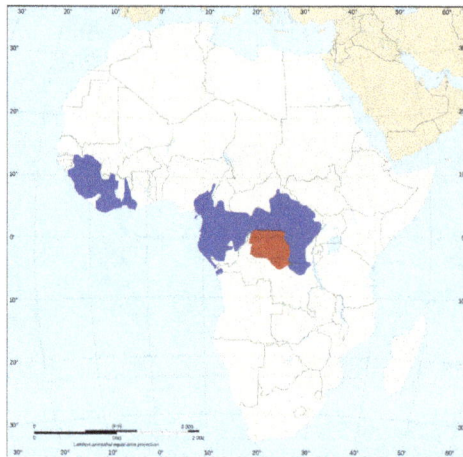

The red shading indicates the range of the bonobo (*Pan paniscus*). The blue shading indicates the range of the Common chimpanzee (*Pan troglodytes*). This is an example of allopatric speciation because they are divided by a natural barrier (the Congo River) and have no habitat in common.

Allopatric speciation (from the ancient Greek *allos*, meaning "other", and *patris*, meaning "fatherland") or geographic speciation is speciation that occurs when biological populations of the same species become vicariant, or isolated from each other to an extent that prevents or interferes with genetic interchange. This can be the result of population dispersal leading to emigration, or by geographical changes such as mountain formation, island formation, or large scale human activities (for example agricultural and civil engineering developments). The vicariant populations then undergo genotypic or phenotypic divergence as: (a) they become subjected to different selective pressures, (b) they independently undergo genetic drift, and (c) different mutations arise in the gene pools of the populations. Allopatric speciation is thought to be the dominant mode of speciation.

The separate populations over time may evolve distinctly different characteristics. If the geographical barriers are later removed, members of the two populations may be unable to successfully mate with each other, at which point, the genetically isolated groups have emerged as different species. Allopatric isolation is a key factor in speciation and a common process by which new species arise. Adaptive radiation, as observed by Charles Darwin in Galapagos finches, is a consequence of allopatric speciation among island populations.

Isolating Mechanisms

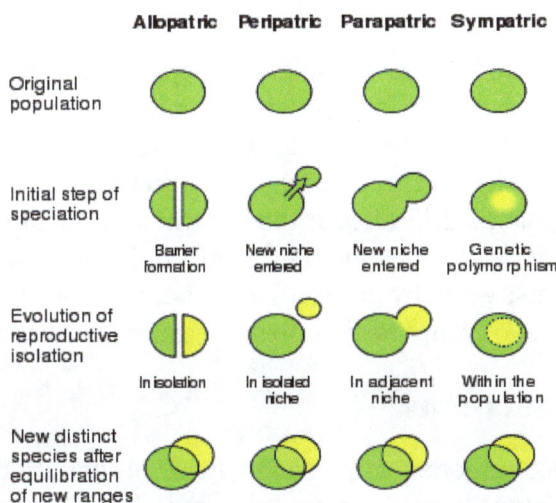

Comparison of allopatric, peripatric, parapatric and sympatric speciation.

Allopatric speciation may occur when a species is subdivided into geographically isolated populations. Such separation commonly is referred to as vicariance. *Allopatric* and *allopatry* are terms from biogeography, referring to organisms whose ranges are entirely separate such that they do not occur in any one place together. If these organisms are closely related (e.g. sister species), such a distribution is usually the result of allopatric speciation. Separation may be attributed to either geological processes or population dispersal.

Geological Processes

Geological processes can fragment a population through such events as emergence of mountain ranges, canyon formation, glacial processes, the formation or destruction of land bridges, or the subsidence of large bodies of water. On a global scale, plate tectonics is a major geological factor that can lead to the separation of populations to result in the distribution of species.

Approximately 50,000 years ago, the Death Valley region of the western United States had a rainy climate which produced an interconnecting system of freshwater rivers and lakes. Climatic changes resulted in a drying trend that has continued for the last 10,000 years. As the lakes and rivers shrank, fish populations became geologically isolated. The few remaining (separated) springs are currently home to a variety of fish, many sharing a close common ancestor; yet each has uniquely adapted to its own particular pool.

The extent to which a geological barrier can effectively isolate a population correlates to the mobility of the organism or its offspring. For example, physical barriers such as canyons may effectively block migration and dispersal of small mammals; however, they have little impact on flying birds or wind-borne seeds.

Population Dispersal

Population dispersal is used to describe migratory events, either in the form of range expansion (natural movement away from parents) or jump dispersal (crossing of barriers), which may lead to genetic isolation. If the smaller population fragment becomes genetically isolated from the parental group, it may be subjected to its own unique mutations, selection forces, and genetic drift effects; thus, it will follow its own evolutionary pathway. Migrations or accidental relocations (such as birds being blown off course) may lead to population fragments; whereby groups merely become separated by distance. Once gene flow between the two groups is disrupted, speciation becomes a possibility.

In Peripheral Populations

When populations become genetically isolated, heritable variations may accumulate so that they become different from the parental population. Given sufficient time, these variations may lead to reproductive isolation.

Portions of a populations that exist along the edges of the parent population's geographic territory have higher likelihood of developing reproductive isolation. Such peripheral populations are likely to possess genes that are different from the parental population. After isolation, the founding population is less likely to represent the gene pool of the parent population. In addition, peripheral isolates are likely to represent a small number of individuals, meaning their gene pool is more susceptible to the effects of genetic drift (random chance). Furthermore, it is likely that the peripheral population will inhabit an environment different from its ancestral gene pool, likely causing it to be subjected to different selective pressures as it colonizes new areas. The outer periphery of a population's habitat tends to be extreme; hence, the reason range expansion is kept in check. For most peripheral isolates, it is more likely that they die off rather than survive and speciate.

Genesis of Reproductive Barriers

Adaptive divergence may occur when a population becomes geographically divided: followed by an accumulation of genetic differences as they adapt to their own unique environments. Reproductive barriers do not evolve as a consequence of external forces that drive populations toward speciation. Rather, the evolution of reproductive isolation, leading to speciation, is generally thought to be an incidental by-product of genetic divergence, particularly adaptive changes that evolve through natural selection in response to different environmental conditions in separate geographic areas.

The *Biological Species Concept,* proposed by Ernst Mayr, in 1942, emphasizes reproductive isolation as the basis of defining a species. The definition states: "A species is defined as a population or group of populations whose members have the potential to interbreed with one another in nature and to produce viable offspring, but cannot produce viable, fertile offspring with members of other species." Mayr, a proponent of allopatric speciation, hypothesized that adaptive genetic changes that accumulate between allopatric populations cause negative epistasis in hybrids, resulting in sterility of the offspring.

If there is considerable genetic and phenotypic change without the loss of the capacity for interbreeding, then such hybridization is simply prevented by the geographical separation of populations. In this case the populations are normally regarded as subspecies.

The frequency of other types of speciation, such as sympatric speciation, parapatric speciation, and heteropatric speciation, is debated. Proponents of peripatric speciation contend that small population size in the peripheral isolate (sometimes referred to as a "splinter population") increases genetic drift, which can be a more powerful force than natural selection in small populations. It deconstructs complex genotypes, allowing the creation of novel gene combinations. Both forms need not be mutually exclusive. In practice, passive isolation or fragmentation as well as active dispersal seem to play a role in many cases of speciation.

Examples

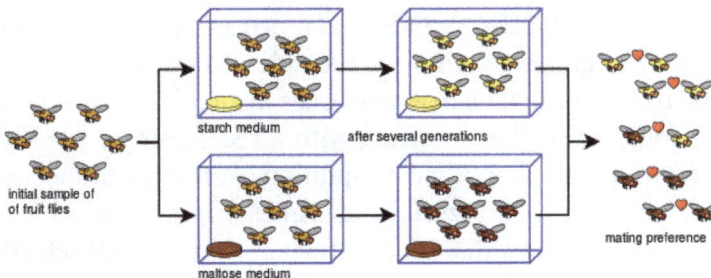

Allopatric speciation among fruit flies

The African elephant has always been regarded as a single species but, because of morphological and DNA differences, some scientists classify them into three subspecies. Researchers at the University of California, San Diego have argued that divergence due to geographical isolation has gone further, and the elephants of West Africa should be regarded as a separate species from both the savanna elephants of Central, Eastern and Southern Africa, and the forest elephants of Central Africa. A similar situation exists with the Asian elephant, which has four distinct living sub-species. The definition of what causes two organisms to be different species is quite ambiguous but the simplest definition is that if the two organisms cannot interbreed they are different species.

Other cases arise where two populations that are quite distinct morphologically, and are native to different continents, have been classified as different species; but when members of one species are introduced into the other's range, they are found to interbreed freely, showing that they were in fact only geographically isolated subspecies. This was found to be the case when the mallard was introduced into New Zealand and interbred freely with the native grey duck, which had been classified as a separate species. It is controversial whether its specific status can now be retained. Mallards interbreed similarly and aggressively with the southern African yellow-billed duck.

Adaptive Radiation

1. Geospiza magnirostris   2. Geospiza fortis
3. Geospiza parvula   4. Certhidea olivacea

Finches from Galapagos Archipelago

Four of the 14 finch species found on the Galápagos Archipelago, are thought to have evolved by an adaptive radiation that diversified their beak shapes to adapt them to different food sources.

In evolutionary biology, adaptive radiation is a process in which organisms diversify rapidly from an ancestral species into a multitude of new forms, particularly when a change in the environment makes new resources available, creates new challenges, or opens new environmental niches. Starting with a recent single ancestor, this process results in the speciation and phenotypic adaptation of an array of species exhibiting different morphological and physiological traits. An example of adaptive radiation would be the avian species of the Hawaiian honeycreepers. Via natural selection, these birds adapted rapidly and converged based on the different environments of the Hawaiian islands.

Much research has been done on adaptive radiation due to its dramatic effects on the diversity of a population. However, more research is needed, especially to fully understand the many factors affecting adaptive radiation. Both empirical and theoretical approaches are helpful, though each has its disadvantages. In order to procure the largest amount of data, empirical and theoretical approaches must be united.

Identification

Four features can be used to identify an adaptive radiation:

1. A common ancestry of component species: specifically a *recent* ancestry. Note that this is not the same as a monophyly in which *all* descendants of a common ancestor are included.

2. A phenotype-environment correlation: a *significant* association between environments and the morphological and physiological traits used to exploit those environments.

3. Trait utility: the performance or fitness advantages of trait values in their corresponding environments.

4. Rapid speciation: presence of one or more *bursts* in the emergence of new species around the time that ecological and phenotypic divergence is underway.

Causes

Innovation

The evolution of a novel feature may permit a clade to diversify by making new areas of morphospace accessible. A classic example is the evolution of a fourth cusp in the mammalian tooth. This trait permits a vast increase in the range of foodstuffs which can be fed on. Evolution of this character has thus increased the number of ecological niches available to mammals. The trait arose a number of times in different groups during the Cenozoic, and in each instance was immediately followed by an adaptive radiation. Birds find other ways to provide for each other, i.e. the evolution of flight opened new avenues for evolution to explore, initiating an adaptive radiation. Other examples include placental gestation (for eutherian mammals), or bipedal locomotion (in hominins).

Examples

Darwin's Finches

One famous example where adaptive radiation is seen is with Darwin's finches. It has been observed by many evolutionary biologists that fragmented landscapes are often a prime location for adaptive radiation to occur. The differences in geography throughout disjointed landscapes such as islands are believed to promote such diversification. Darwin's finches occupy the fragmented landscape of the Galápagos Islands and are diversified into many different species which differ in ecology, song, and morphology, specifically the size and shapes of their beaks. The first obvious explanation for these differences is allopatric speciation, speciation that occurs when populations of the same species become isolated geographically and evolve separately. Because the finches are divided amongst the islands, the birds have been evolving separately for several million years. However, this does not account for the fact that many of the species occur in sympatry, with seven or more species inhabiting the same island. This raises the question as to why these species split when living in the same environment with all the same resources. Petren, Grant, Grant, and Keller proposed that the speciation of the finches occurred in two parts: an initial, easily observable allopatric event followed by a less clear sympatric event. This sympatric event which occurred second was adaptive radiation. This occurred largely to promote specialization upon each island. One major morphological difference among species sharing one island is beak size and shape. Adaptive radiation led to the evolution of different beaks which could access different food and resources. Those with short beaks are better adapted to eating seeds on the ground, those with thin, sharp beaks eat insects, and those with long beaks use their beaks to probe for food inside cacti. With these specializations, seven or more species of finches are able to inhabit the same environments without competition or lack of resources killing several off. In other words, these morphological differences in beak size and shape brought about by adaptive radiation allow the island diversification to persist.

Cichlid Fish

Another famous example is the cichlid fishes in lakes of the East African Rift. The lakes in this area are believed to support and sustain about 2,000 different species of these fish, each with different ecological and morphological characteristics such as body size. Like the Galápagos Islands, these lakes form a fragmented landscape that isolates the cichlid fish from one another, allowing them, and many of the organisms they live with, to evolve separately. The diversity of the lakes is in fact quite extraordinary because the adaptive radiations here are sometimes so young. One thing that has interested scientists about the cichlid fish case is the possibility of convergent evolution, or the evolution of analogous structures independently, driven by similar environmental selection pressures. However, quantitative studies on the convergent evolution of the cichlid fish are limited.

Hawaiian Honeycreepers

Another example of an adaptive radiation would be an endemic species of the Hawaiian Islands. The Hawaiian honeycreepers are a large, highly diverse species which have been part of a vast adaptive radiation, that began as the Hawaiian Islands started to form. The honeycreeper species was shaped by island formation and natural selection. The mechanism by which this adaptive radiation occurred can be described as allopatric speciation via the peripheral isolate model. Each time a new island formed, a dispersal event would occur which would result in new community structures on each island. New selection pressures forced the adaptive radiation of the Hawaiian honeycreepers, as they needed to exploit new resources from the different environments of each island. It has been determined that many of the similar morphologies and behaviors of the Hawaiian Honeycreepers, located on distant islands, are due to convergence of analogous traits caused by similar environments.

Hawaiian Silverswords

Though the most famously recognized cases of adaptive radiation have occurred in animals such as Darwin's finches or the cichlid fish, adaptive radiation certainly occurs in plant species as well. The most famous example of adaptive radiation in plants is quite possibly the Hawaiian silverswords. The Hawaiian silversword alliance consists of twenty-eight species of Hawaiian plants which range from trees to shrubs to vines. This is exceptional diversification as can be seen through the significant morphological differences between each species of the Hawaiian silverswords. With some species, it is virtually impossible to distinguish visually that they were ever part of one species to begin with. These radiations occurred millions of years ago, but through studies over the past few decades, it has been suggested that the rate of speciation and diversification was extremely high. These high rates, as well as the fragmented landscape of the Hawaiian Islands, are key characteristics which point directly to adaptive radiation.

Anolis Lizards

Anolis lizards have been radiating widely in many different environments, including Central and South America, as well as the West Indies and experience great diversity of species just as the finches, cichlid fish, and silverswords. Studies have been done to determine whether radiations occur similarly for these lizards on the mainland as they do on the Caribbean islands or if differences can be observed in how they speciated. It has been observed that in fact, the radiations are very different, and ecological and morphological characteristics that these lizards developed as part of their speciation on the islands and on the mainland are unique. They have clearly evolved differently to the environments they inhabit. The environmental pressures on the Anolis lizards are not the same on the mainland as they are on the islands. There are significantly more predators preying on the Anolis lizards on the mainland. This is but one environmental difference. Other factors play a role in what sort of adaptive radiation will develop. Among the Caribbean islands, a larger perch diameter correlates with longer forelimbs, larger body mass, longer tails, and longer hind limbs. However, on the mainland, a larger perch diameter correlates with shorter tails. This shows that these lizards adapted differently to their environment depending on whether they were located on the mainland or the islands. These differing characteristics reconfirm that most of the adaptive radiation between the mainland and the islands occurred independently. On the islands specifically, species have adapted to certain "microhabitats" in which they require different morphological traits to survive. Irschick (1997) divides these micro-habitats into six groups: "trunk–ground, trunk–crown, grass–bush, crown–giant, twig, and trunk." Different groups of lizards would acquire traits for one of these particular areas that made them more specialized for survival in this microhabitat and not so much in others. Adaptive radiation allows species to acquire the traits they need to survive in these microhabitats and reduce competition to allow the survival of a greater number of organisms as seen in many of the examples before.

Evolution

Evolution is change in the heritable characteristics of biological populations over successive generations. Evolutionary processes give rise to biodiversity at every level of biological organisation, including the levels of species, individual organisms, and molecules.

All life on Earth shares a common ancestor known as the last universal common ancestor (LUCA), which lived approximately 3.5–3.8 billion years ago, although a study in 2015 found "remains of biotic life" from 4.1 billion years ago in ancient rocks in Western Australia. In July 2016, scientists reported identifying a set of 355 genes from the LUCA of all organisms living on Earth.

Repeated formation of new species (speciation), change within species (anagenesis), and loss of species (extinction) throughout the evolutionary history of life on Earth are demonstrated by shared sets of morphological and biochemical traits, including shared DNA sequences. These shared traits are more similar among species that share a more recent common ancestor, and can be used to reconstruct a biological "tree of life" based on evolutionary relationships (phylogenetics), using both existing species and fossils. The fossil record includes a progression from early biogenic graphite, to microbial mat fossils, to fossilized multicellular organisms. Existing patterns of biodiversity have been shaped both by speciation and by extinction. More than 99 percent of all species that ever lived on Earth are estimated to be extinct. Estimates of Earth's current species range from 10 to 14 million, of which about 1.9 million are estimated to have been named and 1.6 million documented in a central database to date. More recently, in May 2016, scientists reported that 1 trillion species are estimated to be on Earth currently with only one-thousandth of one percent described.

In the mid-19th century, Charles Darwin formulated the scientific theory of evolution by natural selection, published in his book *On the Origin of Species* (1859). Evolution by natural selection is a process demonstrated by the observation that more offspring are produced than can possibly survive, along with three facts about populations: 1) traits vary among individuals with respect to morphology, physiology, and behaviour (phenotypic variation), 2) different traits confer different rates of survival and reproduction (differential fitness), and 3) traits can be passed from generation to generation (heritability of fitness). Thus, in successive generations members of a population are replaced by progeny of parents better adapted to survive and reproduce in the biophysical environment in which natural selection takes place. This teleonomy is the quality whereby the process of natural selection creates and preserves traits that are seemingly fitted for the functional roles they perform. The processes by which the changes occur, from one generation to another, are called evolutionary processes or mechanisms. The four most widely recognized evolutionary processes are natural selection (including sexual selection), genetic drift, mutation and gene migration due to genetic admixture. Natural selection and genetic drift sort variation; mutation and gene migration create variation.

Consequences of selection can include meiotic drive (unequal transmission of certain alleles), nonrandom mating and genetic hitchhiking. In the early 20th century the modern evolutionary synthesis integrated classical genetics with Darwin's theory of evolution by natural selection through the discipline of population genetics. The importance of natural selection as a cause of evolution was accepted into other branches of biology. Moreover, previously held notions about evolution, such as orthogenesis, evolutionism, and other beliefs about innate "progress" within the largest-scale trends in evolution, became obsolete. Scientists continue to study various aspects of evolutionary biology by forming and testing hypotheses, constructing mathematical models of theoretical biology and biological theories, using observational data, and performing experiments in both the field and the laboratory.

In terms of practical application, an understanding of evolution has been instrumental to developments in numerous scientific and industrial fields, including agriculture, human and veterinary medicine, and the life sciences in general. Discoveries in evolutionary biology have made a significant impact not just in the traditional branches of biology but also in other academic disciplines, including biological anthropology, and evolutionary psychology. Evolutionary computation, a sub-field of artificial intelligence, involves the application of Darwinian principles to problems in computer science.

History of Evolutionary Thought

In 1842, Charles Darwin penned his first sketch of
On the Origin of Species.

Alfred Russel Wallace

Thomas Robert Malthus

Statistician Ronald Fisher

Classical Times

The proposal that one type of organism could descend from another type goes back to some of the first pre-Socratic Greek philosophers, such as Anaximander and Emped-ocles. Such proposals survived into Roman times. The poet and philosopher Lucretius followed Empedocles in his masterwork *De rerum natura* (*On the Nature of Things*).

Medieval

In contrast to these materialistic views, Aristotelianism considered all natural things as actualisations of fixed natural possibilities, known as forms. This was part of a medieval teleological understanding of nature in which all things have an intended role to play in a divine cosmic order. Variations of this idea became the standard understanding of the Middle Ages and were integrated into Christian learning, but Aristotle did not demand that real types of organisms always correspond one-for-one with exact metaphysical forms and specifically gave examples of how new types of living things could come to be.

Pre-Darwinian

In the 17th century, the new method of modern science rejected the Aristotelian ap-proach. It sought explanations of natural phenomena in terms of physical laws that were the same for all visible things and that did not require the existence of any fixed natural categories or divine cosmic order. However, this new approach was slow to take root in the biological sciences, the last bastion of the concept of fixed natural types. John Ray applied one of the previously more general terms for fixed natural types, "species," to plant and animal types, but he strictly identified each type of living thing as a spe-cies and proposed that each species could be defined by the features that perpetuated themselves generation after generation. The biological classification introduced by Carl Linnaeus in 1735 explicitly recognized the hierarchical nature of species relationships, but still viewed species as fixed according to a divine plan.

Other naturalists of this time speculated on the evolutionary change of species over time according to natural laws. In 1751, Pierre Louis Maupertuis wrote of natural modifications occurring during reproduction and accumulating over many generations to produce new species. Georges-Louis Leclerc, Comte de Buffon suggested that species could degenerate into different organisms, and Erasmus Darwin proposed that all warm-blooded animals could have descended from a single microorganism (or "filament"). The first full-fledged evolutionary scheme was Jean-Baptiste Lamarck's "transmutation" theory of 1809, which envisaged spontaneous generation continually producing simple forms of life that developed greater complexity in parallel lineages with an inherent progressive tendency, and postulated that on a local level these lineages adapted to the environment by inheriting changes caused by their use or disuse in parents. (The latter process was later called Lamarckism.) These ideas were condemned by established naturalists as speculation lacking empirical support. In particular, Georges Cuvier insisted that species were unrelated and fixed, their similarities reflecting divine design for functional needs. In the meantime, Ray's ideas of benevolent design had been developed by William Paley into the *Natural Theology or Evidences of the Existence and Attributes of the Deity* (1802), which proposed complex adaptations as evidence of divine design and which was admired by Charles Darwin.

Darwinian Revolution

The crucial break from the concept of constant typological classes or types in biology came with the theory of evolution through natural selection, which was formulated by Charles Darwin in terms of variable populations. Partly influenced by *An Essay on the Principle of Population* (1798) by Thomas Robert Malthus, Darwin noted that population growth would lead to a "struggle for existence" in which favorable variations prevailed as others perished. In each generation, many offspring fail to survive to an age of reproduction because of limited resources. This could explain the diversity of plants and animals from a common ancestry through the working of natural laws in the same way for all types of organism. Darwin developed his theory of "natural selection" from 1838 onwards and was writing up his "big book" on the subject when Alfred Russel Wallace sent him a version of virtually the same theory in 1858. Their separate papers were presented together at a 1858 meeting of the Linnean Society of London. At the end of 1859, Darwin's publication of his "abstract" as *On the Origin of Species* explained natural selection in detail and in a way that led to an increasingly wide acceptance of Darwin's concepts of evolution at the expense of alternative theories. Thomas Henry Huxley applied Darwin's ideas to humans, using paleontology and comparative anatomy to provide strong evidence that humans and apes shared a common ancestry. Some were disturbed by this since it implied that humans did not have a special place in the universe.

Pangenesis

The mechanisms of reproductive heritability and the origin of new traits remained a mystery. Towards this end, Darwin developed his provisional theory of pangenesis.

In 1865, Gregor Mendel reported that traits were inherited in a predictable manner through the independent assortment and segregation of elements (later known as genes). Mendel's laws of inheritance eventually supplanted most of Darwin's pangenesis theory. August Weismann made the important distinction between germ cells that give rise to gametes (such as sperm and egg cells) and the somatic cells of the body, demonstrating that heredity passes through the germ line only. Hugo de Vries connected Darwin's pangenesis theory to Weismann's germ/soma cell distinction and proposed that Darwin's pangenes were concentrated in the cell nucleus and when expressed they could move into the cytoplasm to change the cells structure. De Vries was also one of the researchers who made Mendel's work well-known, believing that Mendelian traits corresponded to the transfer of heritable variations along the germline. To explain how new variants originate, de Vries developed a mutation theory that led to a temporary rift between those who accepted Darwinian evolution and biometricians who allied with de Vries. In the 1930s, pioneers in the field of population genetics, such as Ronald Fisher, Sewall Wright and J. B. S. Haldane set the foundations of evolution onto a robust statistical philosophy. The false contradiction between Darwin's theory, genetic mutations, and Mendelian inheritance was thus reconciled.

The 'modern Synthesis'

In the 1920s and 1930s the so-called modern synthesis connected natural selection, mutation theory, and Mendelian inheritance into a unified theory that applied generally to any branch of biology. The modern synthesis explained patterns observed across species in populations, through fossil transitions in palaeontology, and complex cellular mechanisms in developmental biology. The publication of the structure of DNA by James Watson and Francis Crick in 1953 demonstrated a physical mechanism for inheritance. Molecular biology improved our understanding of the relationship between genotype and phenotype. Advancements were also made in phylogenetic systematics, mapping the transition of traits into a comparative and testable framework through the publication and use of evolutionary trees. In 1973, evolutionary biologist Theodosius Dobzhansky penned that "nothing in biology makes sense except in the light of evolution," because it has brought to light the relations of what first seemed disjointed facts in natural history into a coherent explanatory body of knowledge that describes and predicts many observable facts about life on this planet.

Further Syntheses

Since then, the modern synthesis has been further extended to explain biological phenomena across the full and integrative scale of the biological hierarchy, from genes to species. This extension, known as evolutionary developmental biology and informally called "evo-devo," emphasises how changes between generations (evolution) acts on patterns of change within individual organisms (development). Since the beginning of the 21st century and in light of discoveries made in recent decades, some biologists have argued for an extended evolutionary synthesis, which would account for the effects of non-genetic inheritance modes, such as epigenetics, parental effects, ecological and cultural inheritance, and evolvability.

Heredity

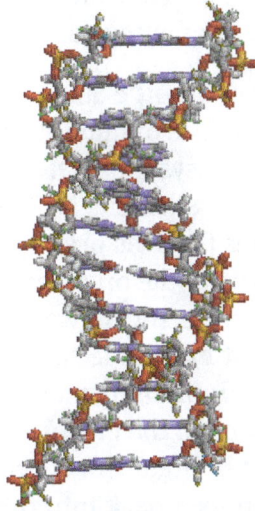

DNA structure. Bases are in the centre, surrounded by phosphate–sugar chains in a double helix.

Evolution in organisms occurs through changes in heritable traits—the inherited characteristics of an organism. In humans, for example, eye colour is an inherited characteristic and an individual might inherit the "brown-eye trait" from one of their parents. Inherited traits are controlled by genes and the complete set of genes within an organism's genome (genetic material) is called its genotype.

The complete set of observable traits that make up the structure and behaviour of an organism is called its phenotype. These traits come from the interaction of its genotype with the environment. As a result, many aspects of an organism's phenotype are not inherited. For example, suntanned skin comes from the interaction between a person's genotype and sunlight; thus, suntans are not passed on to people's children. However, some people tan more easily than others, due to differences in genotypic variation; a striking example are people with the inherited trait of albinism, who do not tan at all and are very sensitive to sunburn.

Heritable traits are passed from one generation to the next via DNA, a molecule that encodes genetic information. DNA is a long biopolymer composed of four types of bases. The sequence of bases along a particular DNA molecule specify the genetic information, in a manner similar to a sequence of letters spelling out a sentence. Before a cell divides, the DNA is copied, so that each of the resulting two cells will inherit the DNA sequence. Portions of a DNA molecule that specify a single functional unit are called genes; different genes have different sequences of bases. Within cells, the long strands of DNA form condensed structures called chromosomes. The specific location of a DNA sequence within a chromosome is known as a locus. If the DNA sequence at a locus varies between individuals, the different forms of this sequence are called alleles. DNA sequences can change through mutations, producing new alleles. If a mutation occurs within a gene, the new allele may affect the trait that the gene controls, altering the phenotype of the organism. However, while this simple correspondence between an allele and a trait works in some cases, most traits are more complex and are controlled by quantitative trait loci (multiple interacting genes).

Recent findings have confirmed important examples of heritable changes that cannot be explained by changes to the sequence of nucleotides in the DNA. These phenomena are classed as epigenetic inheritance systems. DNA methylation marking chromatin, self-sustaining metabolic loops, gene silencing by RNA interference and the three-dimensional conformation of proteins (such as prions) are areas where epigenetic inheritance systems have been discovered at the organismic level. Developmental biologists suggest that complex interactions in genetic networks and communication among cells can lead to heritable variations that may underlay some of the mechanics in developmental plasticity and canalisation. Heritability may also occur at even larger scales. For example, ecological inheritance through the process of niche construction is defined by the regular and repeated activities of organisms in their environment. This generates a legacy of effects that modify and feed back into the selection regime of subsequent generations. Descendants inherit genes plus environmental characteristics generated by the ecological actions of ancestors. Other examples of heritability in evolution that are not under the direct control of genes include the inheritance of cultural traits and symbiogenesis.

Variation

White peppered moth

An individual organism's phenotype results from both its genotype and the influence from the environment it has lived in. A substantial part of the phenotypic variation in a population is caused by genotypic variation. The modern evolutionary synthesis defines evolution as the change over time in this genetic variation. The frequency of one particular allele will become more or less prevalent relative to other forms of that gene. Variation disappears when a new allele reaches the point of fixation—when it either disappears from the population or replaces the ancestral allele entirely.

Natural selection will only cause evolution if there is enough genetic variation in a population. Before the discovery of Mendelian genetics, one common hypothesis was blending inheritance. But with blending inheritance, genetic variance would be rapidly lost, making evolution by natural selection implausible. The Hardy–Weinberg principle provides the solution to how variation is maintained in a population with Mendelian inheritance. The frequencies of alleles (variations in a gene) will remain constant in the absence of selection, mutation, migration and genetic drift.

Black morph in peppered moth evolution

Variation comes from mutations in the genome, reshuffling of genes through sexual reproduction and migration between populations (gene flow). Despite the constant introduction of new variation through mutation and gene flow, most of the genome of a species is identical in all individuals of that species. However, even relatively small differences in genotype can lead to dramatic differences in phenotype: for example, chimpanzees and humans differ in only about 5% of their genomes.

Mutation

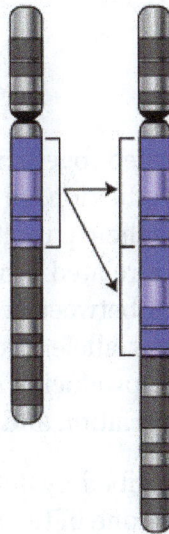

Duplication of part of a chromosome

Mutations are changes in the DNA sequence of a cell's genome. When mutations occur, they may alter the product of a gene, or prevent the gene from functioning, or have no effect. Based on studies in the fly *Drosophila melanogaster*, it has been suggested that if a mutation changes a protein produced by a gene, this will probably be harmful, with about 70% of these mutations having damaging effects, and the remainder being either neutral or weakly beneficial.

Mutations can involve large sections of a chromosome becoming duplicated (usually by genetic recombination), which can introduce extra copies of a gene into a genome. Extra copies of genes are a major source of the raw material needed for new genes to evolve. This is important because most new genes evolve within gene families from pre-existing genes that share common ancestors. For example, the human eye uses four genes to make structures that sense light: three for colour vision and one for night vision; all four are descended from a single ancestral gene.

New genes can be generated from an ancestral gene when a duplicate copy mutates and acquires a new function. This process is easier once a gene has been duplicated because it increases the redundancy of the system; one gene in the pair can acquire a new function while the other copy continues to perform its original function. Other types of mutations can even generate entirely new genes from previously noncoding DNA.

The generation of new genes can also involve small parts of several genes being duplicated, with these fragments then recombining to form new combinations with new functions. When new genes are assembled from shuffling pre-existing parts, domains act as modules with simple independent functions, which can be mixed together to produce new combinations with new and complex functions. For example, polyketide synthases are large enzymes that make antibiotics; they contain up to one hundred independent domains that each catalyse one step in the overall process, like a step in an assembly line.

Sex and Recombination

In asexual organisms, genes are inherited together, or *linked*, as they cannot mix with genes of other organisms during reproduction. In contrast, the offspring of sexual organisms contain random mixtures of their parents' chromosomes that are produced through independent assortment. In a related process called homologous recombination, sexual organisms exchange DNA between two matching chromosomes. Recombination and reassortment do not alter allele frequencies, but instead change which alleles are associated with each other, producing offspring with new combinations of alleles. Sex usually increases genetic variation and may increase the rate of evolution.

The two-fold cost of sex was first described by John Maynard Smith. The first cost is that in sexually dimorphic species only one of the two sexes can bear young. (This cost does not apply to hermaphroditic species, like most plants and many invertebrates.) The second cost is that any individual who reproduces sexually can only pass on 50% of its genes to any individual offspring, with even less passed on as each new generation passes. Yet sexual reproduction is the more common means of reproduction among eukaryotes and multicellular organisms. The Red Queen hypothesis has been used to explain the significance of sexual reproduction as a means to enable continual evolution and adaptation in response to coevolution with other species in an ever-changing environment.

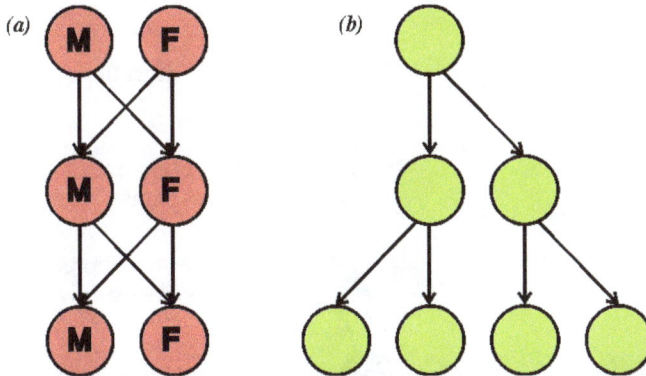

This diagram illustrates the *twofold cost of sex*. If each individual were to contribute to the same number of offspring (two), *(a)* the sexual population remains the same size each generation, where the *(b)* Asexual reproduction population doubles in size each generation.

Gene Flow

Gene flow is the exchange of genes between populations and between species. It can therefore be a source of variation that is new to a population or to a species. Gene flow can be caused by the movement of individuals between separate populations of organisms, as might be caused by the movement of mice between inland and coastal populations, or the movement of pollen between heavy metal tolerant and heavy metal sensitive populations of grasses.

Gene transfer between species includes the formation of hybrid organisms and horizontal gene transfer. Horizontal gene transfer is the transfer of genetic material from one organism to another organism that is not its offspring; this is most common among bacteria. In medicine, this contributes to the spread of antibiotic resistance, as when one bacteria acquires resistance genes it can rapidly transfer them to other species. Horizontal transfer of genes from bacteria to eukaryotes such as the yeast *Saccharomyces cerevisiae* and the adzuki bean weevil *Callosobruchus chinensis* has occurred. An example of larger-scale transfers are the eukaryotic bdelloid rotifers, which have received a range of genes from bacteria, fungi and plants. Viruses can also carry DNA between organisms, allowing transfer of genes even across biological domains.

Large-scale gene transfer has also occurred between the ancestors of eukaryotic cells and bacteria, during the acquisition of chloroplasts and mitochondria. It is possible that eukaryotes themselves originated from horizontal gene transfers between bacteria and archaea.

Mechanisms

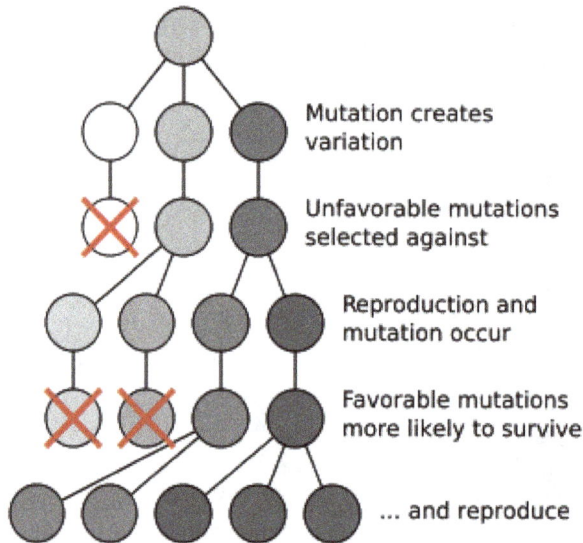

Mutation creates
variation

Unfavorable mutations
selected against

Reproduction and
mutation occur

Favorable mutations
more likely to survive

... and reproduce

Mutation followed by natural selection results in a population with darker colouration.

From a Neo-Darwinian perspective, evolution occurs when there are changes in the frequencies of alleles within a population of interbreeding organisms. For example, the allele for black colour in a population of moths becoming more common. Mechanisms that can lead to changes in allele frequencies include natural selection, genetic drift, genetic hitchhiking, mutation and gene flow.

Natural Selection

Evolution by means of natural selection is the process by which traits that enhance survival and reproduction become more common in successive generations of a population. It has often been called a "self-evident" mechanism because it necessarily follows from three simple facts:

- Variation exists within populations of organisms with respect to morphology, physiology, and behaviour (phenotypic variation).

- Different traits confer different rates of survival and reproduction (differential fitness).

- These traits can be passed from generation to generation (heritability of fitness).

More offspring are produced than can possibly survive, and these conditions produce competition between organisms for survival and reproduction. Consequently, organisms with traits that give them an advantage over their competitors are more likely to pass on their traits to the next generation than those with traits that do not confer an advantage.

The central concept of natural selection is the evolutionary fitness of an organism. Fitness is measured by an organism's ability to survive and reproduce, which determines the size of its genetic contribution to the next generation. However, fitness is not the same as the total number of offspring: instead fitness is indicated by the proportion of subsequent generations that carry an organism's genes. For example, if an organism could survive well and reproduce rapidly, but its offspring were all too small and weak to survive, this organism would make little genetic contribution to future generations and would thus have low fitness.

If an allele increases fitness more than the other alleles of that gene, then with each generation this allele will become more common within the population. These traits are said to be "selected *for*." Examples of traits that can increase fitness are enhanced survival and increased fecundity. Conversely, the lower fitness caused by having a less beneficial or deleterious allele results in this allele becoming rarer—they are "selected *against*." Importantly, the fitness of an allele is not a fixed characteristic; if the environment changes, previously neutral or harmful traits may become beneficial and previously beneficial traits become harmful. However, even if the direction of selection does reverse in this way, traits that were lost in the past may not re-evolve in an identical form.

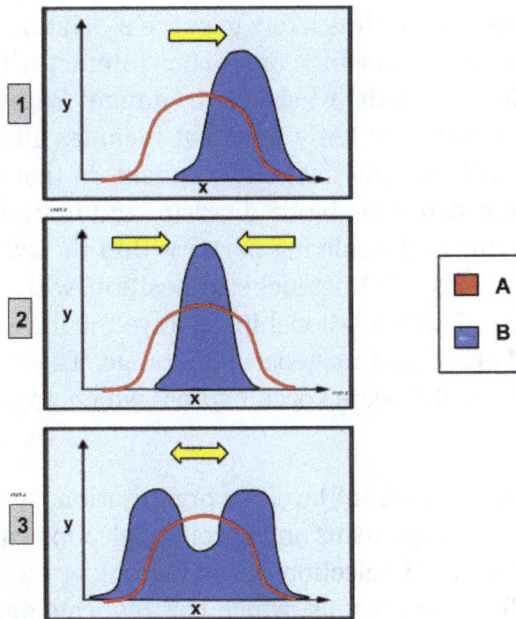

These charts depict the different types of genetic selection. On each graph, the x-axis variable is the type of phenotypic trait and the y-axis variable is the number of organisms. Group A is the original population and Group B is the population after selection.
· Graph 1 shows directional selection, in which a single extreme phenotype is favored.
· Graph 2 depicts stabilizing selection, where the intermediate phenotype is favored over the extreme traits.
· Graph 3 shows disruptive selection, in which the extreme phenotypes are favored over the intermediate.

Natural selection within a population for a trait that can vary across a range of values, such as height, can be categorised into three different types. The first is directional selection, which is a shift in the average value of a trait over time—for example, organisms slowly getting taller. Secondly, disruptive selection is selection for extreme trait values and often results in two different values becoming most common, with selection against the average value. This would be when either short or tall organisms had an advantage, but not those of medium height. Finally, in stabilising selection there is selection against extreme trait values on both ends, which causes a decrease in variance around the average value and less diversity. This would, for example, cause organisms to slowly become all the same height.

A special case of natural selection is sexual selection, which is selection for any trait that increases mating success by increasing the attractiveness of an organism to potential mates. Traits that evolved through sexual selection are particularly prominent among males of several animal species. Although sexually favoured, traits such as cumbersome antlers, mating calls, large body size and bright colours often attract predation, which compromises the survival of individual males. This survival disadvantage is balanced by higher reproductive success in males that show these hard-to-fake, sexually selected traits.

Natural selection most generally makes nature the measure against which individuals and individual traits, are more or less likely to survive. "Nature" in this sense refers to an ecosystem, that is, a system in which organisms interact with every other element, physical as well as biological, in their local environment. Eugene Odum, a founder of ecology, defined an ecosystem as: "Any unit that includes all of the organisms...in a given area interacting with the physical environment so that a flow of energy leads to clearly defined trophic structure, biotic diversity and material cycles (ie: exchange of materials between living and nonliving parts) within the system." Each population within an ecosystem occupies a distinct niche, or position, with distinct relationships to other parts of the system. These relationships involve the life history of the organism, its position in the food chain and its geographic range. This broad understanding of nature enables scientists to delineate specific forces which, together, comprise natural selection.

Natural selection can act at different levels of organisation, such as genes, cells, individual organisms, groups of organisms and species. Selection can act at multiple levels simultaneously. An example of selection occurring below the level of the individual organism are genes called transposons, which can replicate and spread throughout a genome. Selection at a level above the individual, such as group selection, may allow the evolution of cooperation, as discussed below.

Biased Mutation

In addition to being a major source of variation, mutation may also function as a mechanism of evolution when there are different probabilities at the molecular level for different mutations to occur, a process known as mutation bias. If two genotypes, for example one with the nucleotide G and another with the nucleotide A in the same position, have the same fitness, but mutation from G to A happens more often than mutation from A to G, then genotypes with A will tend to evolve. Different insertion vs. deletion mutation biases in different taxa can lead to the evolution of different genome sizes. Developmental or mutational biases have also been observed in morphological evolution. For example, according to the phenotype-first theory of evolution, mutations can eventually cause the genetic assimilation of traits that were previously induced by the environment.

Mutation bias effects are superimposed on other processes. If selection would favor either one out of two mutations, but there is no extra advantage to having both, then the mutation that occurs the most frequently is the one that is most likely to become fixed in a population. Mutations leading to the loss of function of a gene are much more common than mutations that produce a new, fully functional gene. Most loss of function mutations are selected against. But when selection is weak, mutation bias towards loss of function can affect evolution. For example, pigments are no longer useful when animals live in the darkness of caves, and tend to be lost. This kind of loss of function can occur because of mutation bias, and/or because the function had a cost, and once the benefit of the function disappeared, natural selection leads to the loss. Loss of sporulation ability in *Bacillus subtilis* during laboratory evolution appears to have been caused by mutation bias, rather than natural selection against the cost of maintaining sporulation ability. When there is no selection for loss of function, the speed at which loss evolves depends more on the mutation rate than it does on the effective population size, indicating that it is driven more by mutation bias than by genetic drift. In parasitic organisms, mutation bias leads to selection pressures as seen in Ehrlichia. Mutations are biased towards antigenic variants in outer-membrane proteins.

Genetic Drift

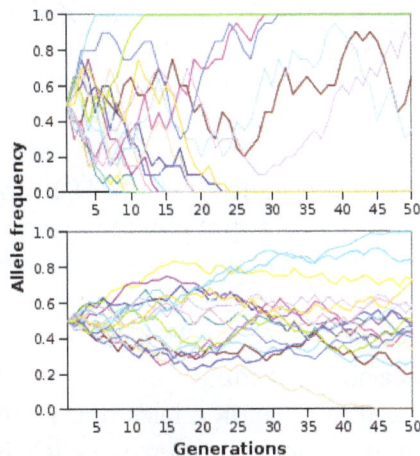

Simulation of genetic drift of 20 unlinked alleles in populations of 10 (top) and 100 (bottom). Drift to fixation is more rapid in the smaller population.

Genetic drift is the change in allele frequency from one generation to the next that occurs because alleles are subject to sampling error. As a result, when selective forces are absent or relatively weak, allele frequencies tend to "drift" upward or downward randomly (in a random walk). This drift halts when an allele eventually becomes fixed, either by disappearing from the population, or replacing the other alleles entirely. Genetic drift may therefore eliminate some alleles from a population due to chance alone. Even in the absence of selective forces, genetic drift can cause two separate populations that began with the same genetic structure to drift apart into two divergent populations with different sets of alleles.

It is usually difficult to measure the relative importance of selection and neutral processes, including drift. The comparative importance of adaptive and non-adaptive forces in driving evolutionary change is an area of current research.

The neutral theory of molecular evolution proposed that most evolutionary changes are the result of the fixation of neutral mutations by genetic drift. Hence, in this model, most genetic changes in a population are the result of constant mutation pressure and genetic drift. This form of the neutral theory is now largely abandoned, since it does not seem to fit the genetic variation seen in nature. However, a more recent and better-supported version of this model is the nearly neutral theory, where a mutation that would be effectively neutral in a small population is not necessarily neutral in a large population. Other alternative theories propose that genetic drift is dwarfed by other stochastic forces in evolution, such as genetic hitchhiking, also known as genetic draft.

The time for a neutral allele to become fixed by genetic drift depends on population size, with fixation occurring more rapidly in smaller populations. The number of individuals in a population is not critical, but instead a measure known as the effective population size. The effective population is usually smaller than the total population since it takes into account factors such as the level of inbreeding and the stage of the lifecycle in which the population is the smallest. The effective population size may not be the same for every gene in the same population.

Genetic Hitchhiking

Recombination allows alleles on the same strand of DNA to become separated. However, the rate of recombination is low (approximately two events per chromosome per generation). As a result, genes close together on a chromosome may not always be shuffled away from each other and genes that are close together tend to be inherited together, a phenomenon known as linkage. This tendency is measured by finding how often two alleles occur together on a single chromosome compared to expectations, which is called their linkage disequilibrium. A set of alleles that is usually inherited in a group is called a haplotype. This can be important when one allele in a particular haplotype is strongly beneficial: natural selection can drive a selective sweep that will also cause the other alleles in the haplotype to become more common in the population; this effect is called genetic hitchhiking or genetic draft. Genetic draft caused by the fact that some neutral genes are genetically linked to others that are under selection can be partially captured by an appropriate effective population size.

Gene Flow

Gene flow involves the exchange of genes between populations and between species. The presence or absence of gene flow fundamentally changes the course of evolution. Due to the complexity of organisms, any two completely isolated populations will eventually evolve genetic incompatibilities through neutral processes, as in the Bateson-Dobzhansky-Muller model, even if both populations remain essentially identical in terms of their adaptation to the environment.

If genetic differentiation between populations develops, gene flow between populations can introduce traits or alleles which are disadvantageous in the local population and this may lead to organisms within these populations evolving mechanisms that prevent mating with genetically distant populations, eventually resulting in the appearance of new species. Thus, exchange of genetic information between individuals is fundamentally important for the development of the biological species concept.

During the development of the modern synthesis, Sewall Wright developed his shifting balance theory, which regarded gene flow between partially isolated populations as an important aspect of adaptive evolution. However, recently there has been substantial criticism of the importance of the shifting balance theory.

Outcomes

Evolution influences every aspect of the form and behaviour of organisms. Most prominent are the specific behavioural and physical adaptations that are the outcome of natural selection. These adaptations increase fitness by aiding activities such as finding food, avoiding predators or attracting mates. Organisms can also respond to selection by cooperating with each other, usually by aiding their relatives or engaging in mutually beneficial symbiosis. In the longer term, evolution produces new species through splitting ancestral populations of organisms into new groups that cannot or will not interbreed.

These outcomes of evolution are distinguished based on time scale as macroevolution versus microevolution. Macroevolution refers to evolution that occurs at or above the level of species, in particular speciation and extinction; whereas microevolution refers to smaller evolutionary changes within a species or population, in particular shifts in gene frequency and adaptation. In general, macroevolution is regarded as the outcome of long periods of microevolution. Thus, the distinction between micro- and macroevolution is not a fundamental one—the difference is simply the time involved. However, in macroevolution, the traits of the entire species may be important. For instance, a large amount of variation among individuals allows a species to rapidly adapt to new habitats, lessening the chance of it going extinct, while a wide geographic range increases the chance of speciation, by making it more likely that part of the population will become isolated. In this sense, microevolution and macroevolution might involve selection at different levels—with microevolution acting on genes and organisms, versus macroevolutionary processes such as species selection acting on entire species and affecting their rates of speciation and extinction.

A common misconception is that evolution has goals, long-term plans, or an innate tendency for "progress," as expressed in beliefs such as orthogenesis and evolutionism; realistically however, evolution has no long-term goal and does not necessarily produce greater complexity. Although complex species have evolved, they occur as a side effect of the overall number of organisms increasing and simple forms of life still remain more common in the biosphere. For example, the overwhelming majority of species are microscopic prokaryotes, which form about half the world's biomass despite their small size, and constitute the vast majority of Earth's biodiversity. Simple organisms have therefore been the dominant form of life on Earth throughout its history and continue to be the main form of life up to the present day, with complex life only appearing more diverse because it is more noticeable. Indeed, the evolution of microorganisms is particularly important to modern evolutionary research, since their rapid reproduction allows the study of experimental evolution and the observation of evolution and adaptation in real time.

Adaptation

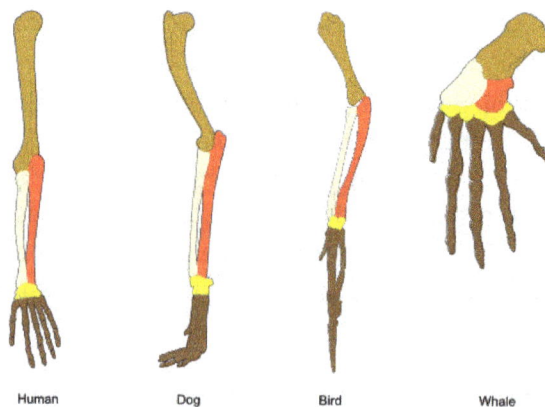

Human   Dog   Bird   Whale

Homologous bones in the limbs of tetrapods. The bones of these animals have the same basic structure, but have been adapted for specific uses.

Adaptation is the process that makes organisms better suited to their habitat. Also, the term adaptation may refer to a trait that is important for an organism's survival. For example, the adaptation of horses' teeth to the grinding of grass. By using the term *adaptation* for the evolutionary process and *adaptive trait* for the product (the bodily part or function), the two senses of the word may be distinguished. Adaptations are produced by natural selection. The following definitions are due to Theodosius Dobzhansky:

1. *Adaptation* is the evolutionary process whereby an organism becomes better able to live in its habitat or habitats.

2. *Adaptedness* is the state of being adapted: the degree to which an organism is able to live and reproduce in a given set of habitats.

3. An *adaptive trait* is an aspect of the developmental pattern of the organism which enables or enhances the probability of that organism surviving and re-producing.

Adaptation may cause either the gain of a new feature, or the loss of an ancestral feature. An example that shows both types of change is bacterial adaptation to antibiotic selection, with genetic changes causing antibiotic resistance by both modifying the target of the drug, or increasing the activity of transporters that pump the drug out of the cell. Other striking examples are the bacteria *Escherichia coli* evolving the ability to use citric acid as a nutrient in a long-term laboratory experiment, *Flavobacterium* evolving a novel enzyme that allows these bacteria to grow on the by-products of nylon manufacturing, and the soil bacterium *Sphingobium* evolving an entirely new metabolic pathway that degrades the synthetic pesticide pentachlorophenol. An interesting but still controversial idea is that some adaptations might increase the ability of organisms to generate genetic diversity and adapt by natural selection (increasing organisms' evolvability).

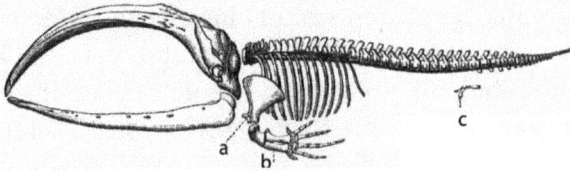

A baleen whale skeleton, *a* and *b* label flipper bones, which were adapted from front leg bones: while *c* indicates vestigial leg bones, suggesting an adaptation from land to sea.

Adaptation occurs through the gradual modification of existing structures. Consequently, structures with similar internal organisation may have different functions in related organisms. This is the result of a single ancestral structure being adapted to function in different ways. The bones within bat wings, for example, are very similar to those in mice feet and primate hands, due to the descent of all these structures from a common mammalian ancestor. However, since all living organisms are related to some extent, even organs that appear to have little or no structural similarity, such as arthropod, squid and vertebrate eyes, or the limbs and wings of arthropods and vertebrates, can depend on a common set of homologous genes that control their assembly and function; this is called deep homology.

During evolution, some structures may lose their original function and become vestigial structures. Such structures may have little or no function in a current species, yet have a clear function in ancestral species, or other closely related species. Examples include pseudogenes, the non-functional remains of eyes in blind cave-dwelling fish, wings in flightless birds, the presence of hip bones in whales and snakes, and sexual traits in organisms that reproduce via asexual reproduction. Examples of vestigial structures in humans include wisdom teeth, the coccyx, the vermiform appendix, and other behavioural vestiges such as goose bumps and primitive reflexes.

However, many traits that appear to be simple adaptations are in fact exaptations: structures originally adapted for one function, but which coincidentally became somewhat useful for some other function in the process. One example is the African lizard *Holaspis guentheri*, which developed an extremely flat head for hiding in crevices, as can be seen by looking at its near relatives. However, in this species, the head has become so flattened that it assists in gliding from tree to tree—an exaptation. Within cells, molecular machines such as the bacterial flagella and protein sorting machinery evolved by the recruitment of several pre-existing proteins that previously had different functions. Another example is the recruitment of enzymes from glycolysis and xenobiotic metabolism to serve as structural proteins called crystallins within the lenses of organisms' eyes.

An area of current investigation in evolutionary developmental biology is the developmental basis of adaptations and exaptations. This research addresses the origin and evolution of embryonic development and how modifications of development and developmental processes produce novel features. These studies have shown that evolution can alter development to produce new structures, such as embryonic bone structures that develop into the jaw in other animals instead forming part of the middle ear in mammals. It is also possible for structures that have been lost in evolution to reappear due to changes in developmental genes, such as a mutation in chickens causing embryos to grow teeth similar to those of crocodiles. It is now becoming clear that most alterations in the form of organisms are due to changes in a small set of conserved genes.

Coevolution

Common garter snake (*Thamnophis sirtalis sirtalis*) has evolved resistance to the defensive substance tetrodotoxin in its amphibian prey.

Interactions between organisms can produce both conflict and cooperation. When the interaction is between pairs of species, such as a pathogen and a host, or a predator and its prey, these species can develop matched sets of adaptations. Here, the evolution of one species causes adaptations in a second species. These changes in the second species then, in turn, cause new adaptations in the first species. This cycle of selection and response is called coevolution. An example is the production of tetrodotoxin in the rough-skinned newt and the evolution of tetrodotoxin resistance in its predator, the common garter snake. In this predator-prey pair, an evolutionary arms race has produced high levels of toxin in the newt and correspondingly high levels of toxin resistance in the snake.

Cooperation

Not all co-evolved interactions between species involve conflict. Many cases of mutually beneficial interactions have evolved. For instance, an extreme cooperation exists between plants and the mycorrhizal fungi that grow on their roots and aid the plant in absorbing nutrients from the soil. This is a reciprocal relationship as the plants provide the fungi with sugars from photosynthesis. Here, the fungi actually grow inside plant cells, allowing them to exchange nutrients with their hosts, while sending signals that suppress the plant immune system.

Coalitions between organisms of the same species have also evolved. An extreme case is the eusociality found in social insects, such as bees, termites and ants, where sterile insects feed and guard the small number of organisms in a colony that are able to reproduce. On an even smaller scale, the somatic cells that make up the body of an animal limit their reproduction so they can maintain a stable organism, which then supports a small number of the animal's germ cells to produce offspring. Here, somatic cells respond to specific signals that instruct them whether to grow, remain as they are, or die. If cells ignore these signals and multiply inappropriately, their uncontrolled growth causes cancer.

Such cooperation within species may have evolved through the process of kin selection, which is where one organism acts to help raise a relative's offspring. This activity is selected for because if the *helping* individual contains alleles which promote the helping activity, it is likely that its kin will *also* contain these alleles and thus those alleles will be passed on. Other processes that may promote cooperation include group selection, where cooperation provides benefits to a group of organisms.

Speciation

Speciation is the process where a species diverges into two or more descendant species.

There are multiple ways to define the concept of "species." The choice of definition is dependent on the particularities of the species concerned. For example, some species concepts apply more readily toward sexually reproducing organisms while others lend themselves better toward asexual organisms. Despite the diversity of various species concepts, these various concepts can be placed into one of three broad philosophical approaches: interbreeding, ecological and phylogenetic. The Biological Species Concept (BSC) is a classic example of the interbreeding approach. Defined by Ernst Mayr in 1942, the BSC states that "species are groups of actually or potentially interbreeding natural populations, which are reproductively isolated from other such groups." Despite its wide and long-term use, the BSC like others is not without controversy, for example because these concepts cannot be applied to prokaryotes, and this is called the species problem. Some researchers have attempted a unifying monistic definition of species, while others adopt a pluralistic approach and suggest that there may be different ways to logically interpret the definition of a species.

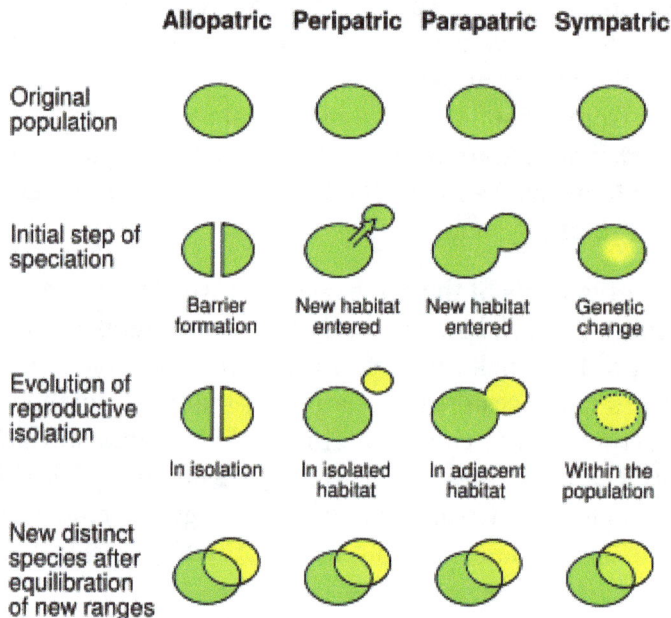

The four mechanisms of speciation

Barriers to reproduction between two diverging sexual populations are required for the populations to become new species. Gene flow may slow this process by spreading the new genetic variants also to the other populations. Depending on how far two species have diverged since their most recent common ancestor, it may still be possible for them to produce offspring, as with horses and donkeys mating to produce mules. Such hybrids are generally infertile. In this case, closely related species may regularly interbreed, but hybrids will be selected against and the species will remain distinct. However, viable hybrids are occasionally formed and these new species can either have properties intermediate between their parent species, or possess a totally new phenotype. The importance of hybridisation in producing new species of animals is unclear, although cases have been seen in many types of animals, with the gray tree frog being a particularly well-studied example.

Speciation has been observed multiple times under both controlled laboratory conditions and in nature. In sexually reproducing organisms, speciation results from reproductive isolation followed by genealogical divergence. There are four mechanisms for speciation. The most common in animals is allopatric speciation, which occurs in populations initially isolated geographically, such as by habitat fragmentation or migration. Selection under these conditions can produce very rapid changes in the appearance and behaviour of organisms. As selection and drift act independently on populations isolated from the rest of their species, separation may eventually produce organisms that cannot interbreed.

The second mechanism of speciation is peripatric speciation, which occurs when small populations of organisms become isolated in a new environment. This differs from allopatric speciation in that the isolated populations are numerically much smaller than the parental population. Here, the founder effect causes rapid speciation after an increase in inbreeding increases selection on homozygotes, leading to rapid genetic change.

The third mechanism of speciation is parapatric speciation. This is similar to peripatric speciation in that a small population enters a new habitat, but differs in that there is no physical separation between these two populations. Instead, speciation results from the evolution of mechanisms that reduce gene flow between the two populations. Generally this occurs when there has been a drastic change in the environment within the parental species' habitat. One example is the grass *Anthoxanthum odoratum*, which can undergo parapatric speciation in response to localised metal pollution from mines. Here, plants evolve that have resistance to high levels of metals in the soil. Selection against interbreeding with the metal-sensitive parental population produced a gradual change in the flowering time of the metal-resistant plants, which eventually produced complete reproductive isolation. Selection against hybrids between the two populations may cause *reinforcement*, which is the evolution of traits that promote mating within a species, as well as character displacement, which is when two species become more distinct in appearance.

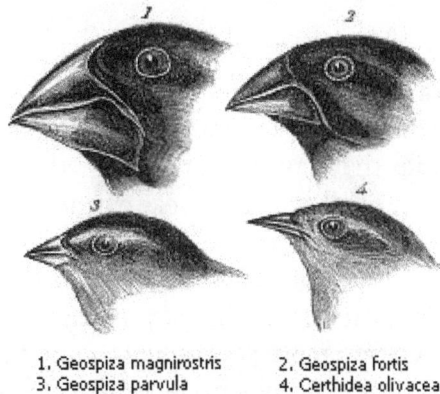

1. Geospiza magnirostris   2. Geospiza fortis
3. Geospiza parvula   4. Certhidea olivacea

Finches from Galapagos Archipelago

Geographical isolation of finches on the Galápagos Islands produced over a dozen new species.

Finally, in sympatric speciation species diverge without geographic isolation or changes in habitat. This form is rare since even a small amount of gene flow may remove genetic differences between parts of a population. Generally, sympatric speciation in animals requires the evolution of both genetic differences and non-random mating, to allow reproductive isolation to evolve.

One type of sympatric speciation involves crossbreeding of two related species to produce a new hybrid species. This is not common in animals as animal hybrids are usually sterile. This is because during meiosis the homologous chromosomes from each parent are from different species and cannot successfully pair. However, it is more common in plants because plants often double their number of chromosomes, to form polyploids. This allows the chromosomes from each parental species to form matching pairs during meiosis, since each parent's chromosomes are represented by a pair already. An example of such a speciation event is when the plant species *Arabidopsis thaliana* and *Arabidopsis arenosa* crossbred to give the new species *Arabidopsis suecica*. This happened about 20,000 years ago, and the speciation process has been repeated in the laboratory, which allows the study of the genetic mechanisms involved in this process. Indeed, chromosome doubling within a species may be a common cause of reproductive isolation, as half the doubled chromosomes will be unmatched when breeding with undoubled organisms.

Speciation events are important in the theory of punctuated equilibrium, which accounts for the pattern in the fossil record of short "bursts" of evolution interspersed with relatively long periods of stasis, where species remain relatively unchanged. In this theory, speciation and rapid evolution are linked, with natural selection and genetic drift acting most strongly on organisms undergoing speciation in novel habitats or small populations. As a result, the periods of stasis in the fossil record correspond to the parental population and the organisms undergoing speciation and rapid evolution are found in small populations or geographically restricted habitats and therefore rarely being preserved as fossils.

Extinction

Extinction is the disappearance of an entire species. Extinction is not an unusual event, as species regularly appear through speciation and disappear through extinction. Nearly all animal and plant species that have lived on Earth are now extinct, and extinction appears to be the ultimate fate of all species. These extinctions have happened continuously throughout the history of life, although the rate of extinction spikes in occasional mass extinction events. The Cretaceous–Paleogene extinction event, during which the non-avian dinosaurs became extinct, is the most well-known, but the earlier Permian–Triassic extinction event was even more severe, with approximately 96% of all marine species driven to extinction. The Holocene extinction event is an ongoing mass extinction associated with humanity's expansion across the globe over the past few thousand years. Present-day extinction rates are 100–1000 times greater than the background rate and up to 30% of current species may be extinct by the mid 21st century. Human activities are now the primary cause of the ongoing extinction event; global warming may further accelerate it in the future.

Tyrannosaurus rex. Non-avian dinosaurs died out in the Cretaceous–Paleogene extinction event at the end of the Cretaceous period.

The role of extinction in evolution is not very well understood and may depend on which type of extinction is considered. The causes of the continuous "low-level" extinction events, which form the majority of extinctions, may be the result of competition between species for limited resources (the competitive exclusion principle). If one species can out-compete another, this could produce species selection, with the fitter species surviving and the other species being driven to extinction. The intermittent mass extinctions are also important, but instead of acting as a selective force, they drastically reduce diversity in a nonspecific manner and promote bursts of rapid evolution and speciation in survivors.

Evolutionary History of Life

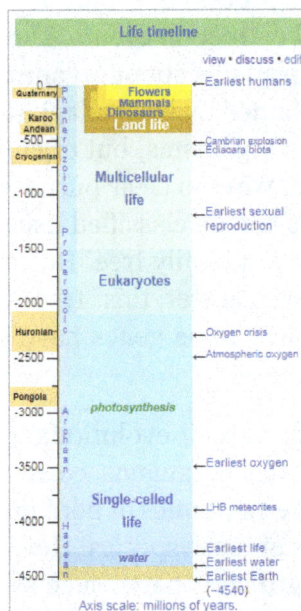

Origin of Life

The Earth is about 4.54 billion years old. The earliest undisputed evidence of life on Earth dates from at least 3.5 billion years ago, during the Eoarchean Era after a geological crust started to solidify following the earlier molten Hadean Eon. Microbial mat fossils have been found in 3.48 billion-year-old sandstone in Western Australia. Other early physical evidence of a biogenic substance is graphite in 3.7 billion-year-old metasedimentary rocks discovered in Western Greenland as well as "remains of biotic life" found in 4.1 billion-year-old rocks in Western Australia. According to one of the researchers, "If life arose relatively quickly on Earth ... then it could be common in the universe."

More than 99 percent of all species, amounting to over five billion species, that ever lived on Earth are estimated to be extinct. Estimates on the number of Earth's current species range from 10 million to 14 million, of which about 1.9 million are estimated to have been named and 1.6 million documented in a central database to date, leaving at least 80 percent not yet described.

Highly energetic chemistry is thought to have produced a self-replicating molecule around 4 billion years ago, and half a billion years later the last common ancestor of all life existed. The current scientific consensus is that the complex biochemistry that makes up life came from simpler chemical reactions. The beginning of life may have included self-replicating molecules such as RNA and the assembly of simple cells.

Common Descent

All organisms on Earth are descended from a common ancestor or ancestral gene pool. Current species are a stage in the process of evolution, with their diversity the product of a long series of speciation and extinction events. The common descent of organisms was first deduced from four simple facts about organisms: First, they have geographic distributions that cannot be explained by local adaptation. Second, the diversity of life is not a set of completely unique organisms, but organisms that share morphological similarities. Third, vestigial traits with no clear purpose resemble functional ancestral traits and finally, that organisms can be classified using these similarities into a hierarchy of nested groups—similar to a family tree. However, modern research has suggested that, due to horizontal gene transfer, this "tree of life" may be more complicated than a simple branching tree since some genes have spread independently between distantly related species.

Past species have also left records of their evolutionary history. Fossils, along with the comparative anatomy of present-day organisms, constitute the morphological, or anatomical, record. By comparing the anatomies of both modern and extinct species, paleontologists can infer the lineages of those species. However, this approach is most successful for organisms that had hard body parts, such as shells, bones or teeth. Further, as prokaryotes such as bacteria and archaea share a limited set of common morphologies, their fossils do not provide information on their ancestry.

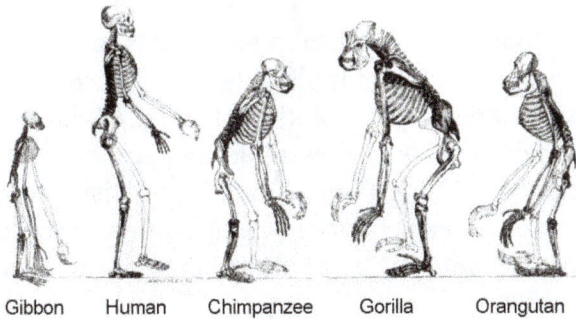

Gibbon   Human   Chimpanzee   Gorilla   Orangutan

The hominoids are descendants of a common ancestor.

More recently, evidence for common descent has come from the study of biochemical similarities between organisms. For example, all living cells use the same basic set of nucleotides and amino acids. The development of molecular genetics has revealed the record of evolution left in organisms' genomes: dating when species diverged through the molecular clock produced by mutations. For example, these DNA sequence comparisons have revealed that humans and chimpanzees share 98% of their genomes and analysing the few areas where they differ helps shed light on when the common ancestor of these species existed.

Evolution of Life

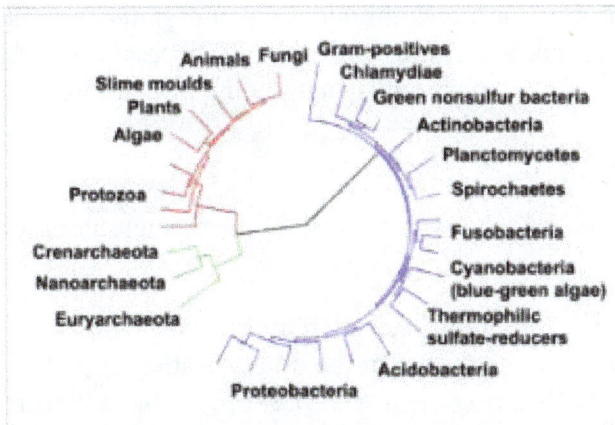

Evolutionary tree showing the divergence of modern species from their common ancestor in the centre. The three domains are coloured, with bacteria blue, archaea green and eukaryotes red.

Prokaryotes inhabited the Earth from approximately 3–4 billion years ago. No obvious changes in morphology or cellular organisation occurred in these organisms over the next few billion years. The eukaryotic cells emerged between 1.6–2.7 billion years ago. The next major change in cell structure came when bacteria were engulfed by eukaryotic cells, in a cooperative association called endosymbiosis. The engulfed bacteria and the host cell then underwent coevolution, with the bacteria evolving into either mitochondria or hydrogenosomes. Another engulfment of cyanobacterial-like organisms led to the formation of chloroplasts in algae and plants.

The history of life was that of the unicellular eukaryotes, prokaryotes and archaea until about 610 million years ago when multicellular organisms began to appear in the oceans in the Ediacaran period. The evolution of multicellularity occurred in multiple independent events, in organisms as diverse as sponges, brown algae, cyanobacteria, slime moulds and myxobacteria. In January 2016, scientists reported that, about 800 million years ago, a minor genetic change in a single molecule called GK-PID may have allowed organisms to go from a single cell organism to one of many cells.

Soon after the emergence of these first multicellular organisms, a remarkable amount of biological diversity appeared over approximately 10 million years, in an event called the Cambrian explosion. Here, the majority of types of modern animals appeared in the fossil record, as well as unique lineages that subsequently became extinct. Various triggers for the Cambrian explosion have been proposed, including the accumulation of oxygen in the atmosphere from photosynthesis.

About 500 million years ago, plants and fungi colonised the land and were soon followed by arthropods and other animals. Insects were particularly successful and even today make up the majority of animal species. Amphibians first appeared around 364 million years ago, followed by early amniotes and birds around 155 million years ago (both from "reptile"-like lineages), mammals around 129 million years ago, homininae around 10 million years ago and modern humans around 250,000 years ago. However, despite the evolution of these large animals, smaller organisms similar to the types that evolved early in this process continue to be highly successful and dominate the Earth, with the majority of both biomass and species being prokaryotes.

Applications

Concepts and models used in evolutionary biology, such as natural selection, have many applications.

Artificial selection is the intentional selection of traits in a population of organisms. This has been used for thousands of years in the domestication of plants and animals. More recently, such selection has become a vital part of genetic engineering, with selectable markers such as antibiotic resistance genes being used to manipulate DNA. Proteins with valuable properties have evolved by repeated rounds of mutation and selection (for example modified enzymes and new antibodies) in a process called directed evolution.

Understanding the changes that have occurred during an organism's evolution can reveal the genes needed to construct parts of the body, genes which may be involved in human genetic disorders. For example, the Mexican tetra is an albino cavefish that lost its eyesight during evolution. Breeding together different populations of this blind fish produced some offspring with functional eyes, since different mutations had occurred in the isolated populations that had evolved in different caves. This helped identify genes required for vision and pigmentation.

Many human diseases are not static phenomena, but capable of evolution. Viruses, bacteria, fungi and cancers evolve to be resistant to host immune defences, as well as pharmaceutical drugs. These same problems occur in agriculture with pesticide and herbicide resistance. It is possible that we are facing the end of the effective life of most of available antibiotics and predicting the evolution and evolvability of our pathogens and devising strategies to slow or circumvent it is requiring deeper knowledge of the complex forces driving evolution at the molecular level.

In computer science, simulations of evolution using evolutionary algorithms and artificial life started in the 1960s and were extended with simulation of artificial selection. Artificial evolution became a widely recognised optimisation method as a result of the work of Ingo Rechenberg in the 1960s. He used evolution strategies to solve complex engineering problems. Genetic algorithms in particular became popular through the writing of John Henry Holland. Practical applications also include automatic evolution of computer programmes. Evolutionary algorithms are now used to solve multi-dimensional problems more efficiently than software produced by human designers and also to optimise the design of systems.

Social and Cultural Responses

As evolution became widely accepted in the 1870s, caricatures of Charles Darwin with an ape or monkey body symbolised evolution.

In the 19th century, particularly after the publication of *On the Origin of Species* in 1859, the idea that life had evolved was an active source of academic debate centred on the philosophical, social and religious implications of evolution. Today, the modern evolutionary synthesis is accepted by a vast majority of scientists. However, evolution remains a contentious concept for some theists.

While various religions and denominations have reconciled their beliefs with evolution through concepts such as theistic evolution, there are creationists who believe that evolution is contradicted by the creation myths found in their religions and who raise various objections to evolution. As had been demonstrated by responses to the publication of *Vestiges of the Natural History of Creation* in 1844, the most controversial aspect of evolutionary biology is the implication of human evolution that humans share common ancestry with apes and that the mental and moral faculties of humanity have the same types of natural causes as other inherited traits in animals. In some countries, notably the United States, these tensions between science and religion have fuelled the current creation–evolution controversy, a religious conflict focusing on politics and public education. While other scientific fields such as cosmology and Earth science also conflict with literal interpretations of many religious texts, evolutionary biology experiences significantly more opposition from religious literalists.

The teaching of evolution in American secondary school biology classes was uncommon in most of the first half of the 20th century. The Scopes Trial decision of 1925 caused the subject to become very rare in American secondary biology textbooks for a generation, but it was gradually re-introduced later and became legally protected with the 1968 *Epperson v. Arkansas* decision. Since then, the competing religious belief of creationism was legally disallowed in secondary school curricula in various decisions in the 1970s and 1980s, but it returned in pseudoscientific form as intelligent design (ID), to be excluded once again in the 2005 *Kitzmiller v. Dover Area School District* case.

Extinction

In biology and ecology, extinction is the end of an organism or of a group of organisms (taxon), normally a species. The moment of extinction is generally considered to be the death of the last individual of the species, although the capacity to breed and recover may have been lost before this point. Because a species' potential range may be very large, determining this moment is difficult, and is usually done retrospectively. This difficulty leads to phenomena such as Lazarus taxa, where a species presumed extinct abruptly "reappears" (typically in the fossil record) after a period of apparent absence.

More than 99 percent of all species, amounting to over five billion species, that ever lived on Earth are estimated to be extinct. Estimates on the number of Earth's current species range from 10 million to 14 million, of which about 1.2 million have been documented and over 86 percent have not yet been described. More recently, in May 2016, scientists reported that 1 trillion species are estimated to be on Earth currently with only one-thousandth of one percent described.

Through evolution, species arise through the process of speciation—where new varieties of organisms arise and thrive when they are able to find and exploit an ecological niche—and species become extinct when they are no longer able to survive in changing conditions or against superior competition. The relationship between animals and their ecological niches has been firmly established. A typical species becomes extinct within 10 million years of its first appearance, although some species, called living fossils, survive with virtually no morphological change for hundreds of millions of years.

Mass extinctions are relatively rare events; however, isolated extinctions are quite common. Only recently have extinctions been recorded and scientists have become alarmed at the current high rate of extinctions. Most species that become extinct are never scientifically documented. Some scientists estimate that up to half of presently existing plant and animal species may become extinct by 2100.

A dagger symbol (†) next to a species name is often used to indicate its extinction.

Definition

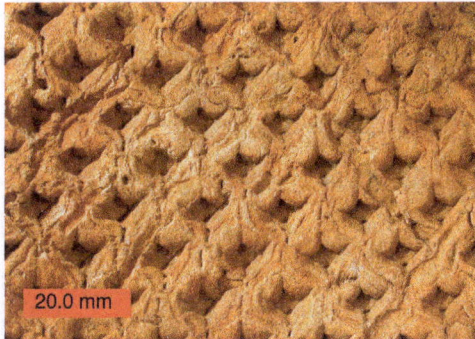

20.0 mm

External mold of the extinct *Lepidodendron* from the Upper Carboniferous of Ohio

A species is extinct when the last existing member dies. Extinction therefore becomes a certainty when there are no surviving individuals that can reproduce and create a new generation. A species may become functionally extinct when only a handful of individuals survive, which cannot reproduce due to poor health, age, sparse distribution over a large range, a lack of individuals of both sexes (in sexually reproducing species), or other reasons.

Pinpointing the extinction (or pseudoextinction) of a species requires a clear definition of that species. If it is to be declared extinct, the species in question must be uniquely distinguishable from any ancestor or daughter species, and from any other closely related species. Extinction of a species (or replacement by a daughter species) plays a key role in the punctuated equilibrium hypothesis of Stephen Jay Gould and Niles Eldredge.

Skeleton of various extinct dinosaurs; some other dinosaur lineages still flourish in the form of birds

In ecology, *extinction* is often used informally to refer to local extinction, in which a species ceases to exist in the chosen area of study, but may still exist elsewhere. This phenomenon is also known as extirpation. Local extinctions may be followed by a replacement of the species taken from other locations; wolf reintroduction is an example of this. Species which are not extinct are termed extant. Those that are extant but threatened by extinction are referred to as threatened or endangered species.

The dodo of Mauritius, shown here in a 1626 illustration by Roelant Savery, is an often-cited example of modern extinction

Currently an important aspect of extinction is human attempts to preserve critically endangered species. These are reflected by the creation of the conservation status "extinct in the wild" (EW). Species listed under this status by the International Union for Conservation of Nature (IUCN) are not known to have any living specimens in the wild, and are maintained only in zoos or other artificial environments. Some of these species are functionally extinct, as they are no longer part of their natural habitat and it is unlikely the species will ever be restored to the wild. When possible, modern zoological institutions try to maintain a viable population for species preservation and possible future reintroduction to the wild, through use of carefully planned breeding programs.

The extinction of one species' wild population can have knock-on effects, causing further extinctions. These are also called "chains of extinction". This is especially common with extinction of keystone species.

Pseudoextinction

Extinction of a parent species where daughter species or subspecies are still extant is called pseudoextinction or phyletic extinction. Effectively, the old taxon vanishes, transformed (anagenesis) into a successor, or split into more than one (cladogenesis).

Pseudoextinction is difficult to demonstrate unless one has a strong chain of evidence linking a living species to members of a pre-existing species. For example, it is sometimes claimed that the extinct *Hyracotherium*, which was an early horse that shares a common ancestor with the modern horse, is pseudoextinct, rather than extinct, because there are several extant species of *Equus*, including zebra and donkey. However, as fossil species typically leave no genetic material behind, one cannot say whether *Hyracotherium* evolved into more modern horse species or merely evolved from a common ancestor with modern horses. Pseudoextinction is much easier to demonstrate for larger taxonomic groups.

Lazarus Taxa

The coelacanth, a fish related to lungfish and tetrapods, was considered to have been extinct since the end of the Cretaceous Period until 1938 when a specimen was found, off the Chalumna River (now Tyolomnqa) on the east coast of South Africa. Museum curator Marjorie Courtenay-Latimer discovered the fish among the catch of a local angler, Captain Hendrick Goosen, on December 23, 1938. A local chemistry professor, JLB Smith, confirmed the fish's importance with a famous cable: "MOST IMPORTANT PRESERVE SKELETON AND GILLS = FISH DESCRIBED".

Far more recent possible or presumed extinctions of species which may turn out still to exist include the thylacine, or Tasmanian tiger (*Thylacinus cynocephalus*), the last known example of which died in Hobart Zoo in Tasmania in 1936; the Japanese wolf (*Canis lupus hodophilax*), last sighted over 100 years ago; the ivory-billed woodpecker (*Campephilus principalis*), last sighted for certain in 1944; and the slender-billed curlew (*Numenius tenuirostris*), not seen since 2007.

Causes

As long as species have been evolving, species have been going extinct. It is estimated that over 99.9% of all species that ever lived are extinct. The average lifespan of a species is 1–10 million years, although this varies widely between taxa. There are a variety of causes that can contribute directly or indirectly to the extinction of a species or group of species. "Just as each species is unique", write Beverly and Stephen C. Stearns, "so is each extinction ... the causes for each are varied—some subtle and complex, others obvious and simple". Most simply, any species that cannot survive and reproduce in its environment and cannot move to a new environment where it can do so, dies out and becomes extinct. Extinction of a species may come suddenly when an otherwise healthy species is wiped out completely, as when toxic pollution renders its entire habitat un-liveable; or may occur gradually over thousands or millions of years, such as when a species gradually loses out in competition for food to better adapted competitors. Extinction may occur a long time after the events that set it in motion, a phenomenon known as extinction debt.

The passenger pigeon, one of hundreds of species of extinct birds, was hunted to extinction over the course of a few decades

Assessing the relative importance of genetic factors compared to environmental ones as the causes of extinction has been compared to the debate on nature and nurture. The question of whether more extinctions in the fossil record have been caused by evolution or by catastrophe is a subject of discussion; Mark Newman, the author of *Modeling Extinction*, argues for a mathematical model that falls between the two positions. By contrast, conservation biology uses the extinction vortex model to classify extinctions by cause. When concerns about human extinction have been raised, for example in Sir Martin Rees' 2003 book *Our Final Hour*, those concerns lie with the effects of climate change or technological disaster.

Currently, environmental groups and some governments are concerned with the ex-tinction of species caused by humanity, and they try to prevent further extinctions through a variety of conservation programs. Humans can cause extinction of a species through overharvesting, pollution, habitat destruction, introduction of invasive species (such as new predators and food competitors), overhunting, and other influences. Ex-plosive, unsustainable human population growth is an essential cause of the extinction crisis. According to the International Union for Conservation of Nature (IUCN), 784 extinctions have been recorded since the year 1500, the arbitrary date selected to define "recent" extinctions, up to the year 2004; with many more likely to have gone unnoticed. Several species have also been listed as extinct since 2004.

Genetics and Demographic Phenomena

If adaptation increasing population fitness is slower than environmental degradation plus the accumulation of slightly deleterious mutations, then a population will go extinct. Smaller populations have fewer beneficial mutations entering the population each generation, slowing adaptation. It is also easier for slightly deleterious mutations to fix in small populations; the resulting positive feedback loop between small population size and low fitness can cause mutational meltdown.

Limited geographic range is the most important determinant of genus extinction at background rates but becomes increasingly irrelevant as mass extinction arises. Limited geographic range is a cause both of small population size and of greater vulnerability to local environnmental catastrophes.

Extinction rates can be affected not just by population size, but by any factor that affects evolvability, including balancing selection, cryptic genetic variation, phenotypic plasticity, and robustness. A diverse or deep gene pool gives a population a higher chance in the short term of surviving an adverse change in conditions. Effects that cause or reward a loss in genetic diversity can increase the chances of extinction of a species. Population bottlenecks can dramatically reduce genetic diversity by severely limiting the number of reproducing individuals and make inbreeding more frequent.

Genetic Pollution

Purebred wild species evolved to a specific ecology can be threatened with extinction through the process of genetic pollution—i.e., uncontrolled hybridization, introgression genetic swamping which leads to homogenization or out-competition from the introduced (or hybrid) species. Endemic populations can face such extinctions when new populations are imported or selectively bred by people, or when habitat modification brings previously isolated species into contact. Extinction is likeliest for rare species coming into contact with more abundant ones; interbreeding can swamp the rarer gene pool and create hybrids, depleting the purebred gene pool (for example, the endangered wild water buffalo is most threatened with extinction by genetic pollution from the abundant domestic water buffalo). Such extinctions are not always apparent from morphological (non-genetic) observations. Some degree of gene flow is a normal evolutionarily process, nevertheless, hybridization (with or without introgression) threatens rare species' existence.

The gene pool of a species or a population is the variety of genetic information in its living members. A large gene pool (extensive genetic diversity) is associated with robust populations that can survive bouts of intense selection. Meanwhile, low genetic diversity reduces the range of adaptions possible. Replacing native with alien genes narrows genetic diversity within the original population, thereby increasing the chance of extinction.

Scorched land resulting from slash-and-burn agriculture

Habitat Degradation

Habitat degradation is currently the main anthropogenic cause of species extinctions. The main cause of habitat degradation worldwide is agriculture, with urban sprawl, logging, mining and some fishing practices close behind. The degradation of a species' habitat may alter the fitness landscape to such an extent that the species is no longer able to survive and becomes extinct. This may occur by direct effects, such as the environment becoming toxic, or indirectly, by limiting a species' ability to compete effectively for diminished resources or against new competitor species.

Habitat degradation through toxicity can kill off a species very rapidly, by killing all living members through contamination or sterilizing them. It can also occur over longer periods at lower toxicity levels by affecting life span, reproductive capacity, or competitiveness.

Habitat degradation can also take the form of a physical destruction of niche habitats. The widespread destruction of tropical rainforests and replacement with open pastureland is widely cited as an example of this; elimination of the dense forest eliminated the infrastructure needed by many species to survive. For example, a fern that depends on dense shade for protection from direct sunlight can no longer survive without forest to shelter it. Another example is the destruction of ocean floors by bottom trawling.

Diminished resources or introduction of new competitor species also often accompany habitat degradation. Global warming has allowed some species to expand their range, bringing unwelcome competition to other species that previously occupied that area. Sometimes these new competitors are predators and directly affect prey species, while at other times they may merely outcompete vulnerable species for limited resources. Vital resources including water and food can also be limited during habitat degradation, leading to extinction.

The golden toad was last seen on May 15, 1989. Decline in amphibian populations is ongoing worldwide

Predation, Competition, and Disease

In the natural course of events, species become extinct for a number of reasons, including but not limited to: extinction of a necessary host, prey or pollinator, inter-species competition, inability to deal with evolving diseases and changing environmental conditions (particularly sudden changes) which can act to introduce novel predators, or to remove prey. Recently in geological time, humans have become an additional cause of extinction (many people would say premature extinction) of some species, either as a new mega-predator or by transporting animals and plants from one part of the world to another. Such introductions have been occurring for thousands of years, sometimes intentionally (e.g. livestock released by sailors on islands as a future source of food) and sometimes accidentally (e.g. rats escaping from boats). In most cases, the introductions are unsuccessful, but when an invasive alien species does become established, the consequences can be catastrophic. Invasive alien species can affect native species directly by eating them, competing with them, and introducing pathogens or parasites that sicken or kill them; or indirectly by destroying or degrading their habitat. Human populations may themselves act as invasive predators. According to the "overkill hypothesis", the swift extinction of the megafauna in areas such as Australia (40,000 years before present), North and South America (12,000 years before present), Madagascar, Hawaii (300–1000 CE), and New Zealand (1300–1500 CE), resulted from the sudden introduction of human beings to environments full of animals that had never seen them before, and were therefore completely unadapted to their predation techniques.

Coextinction

Coextinction refers to the loss of a species due to the extinction of another; for example, the extinction of parasitic insects following the loss of their hosts. Coextinction can also occur when a species loses its pollinator, or to predators in a food chain who lose their prey. "Species coextinction is a manifestation of the interconnectedness of organisms in complex ecosystems ... While coextinction may not be the most important cause of species extinctions, it is certainly an insidious one". Coextinction is especially common when a keystone species goes extinct. Models suggest that co-extinction is the most common form of biodiversity loss. There may be a cascade of coextinction across the trophic levels. Such effects are most severe in mutualistic and parasitic relationships. An example of coextinction is the Haast's eagle and the moa: the Haast's eagle was a predator that became extinct because its food source became extinct. The moa were several species of flightless birds that were a food source for the Haast's eagle.

The large Haast's eagle and moa from New Zealand

Climate Change

Extinction as a result of climate change has been confirmed by fossil studies. Particularly, the extinction of amphibians during the Carboniferous Rainforest Collapse, 305 million years ago. A 2003 review across 14 biodiversity research centers predicted that, because of climate change, 15–37% of land species would be "committed to extinction" by 2050. The ecologically rich areas that would potentially suffer the heaviest losses include the Cape Floristic Region, and the Caribbean Basin. These areas might see a doubling of present carbon dioxide levels and rising temperatures that could eliminate 56,000 plant and 3,700 animal species.

Mass Extinctions

There have been at least five mass extinctions in the history of life on earth, and four in the last 350 million years in which many species have disappeared in a relatively short period of geological time. A massive eruptive event is considered to be one likely cause of the "Permian–Triassic extinction event" about 250 million years ago, which is estimated to have killed 90% of species then existing. There is also evidence to suggest that this event was preceded by another mass extinction, known as Olson's Extinction. The Cretaceous–Paleogene extinction event (K-Pg) occurred 66 million years ago, at the end of the Cretaceous period, and is best known for having wiped out non-avian dinosaurs, among many other species.

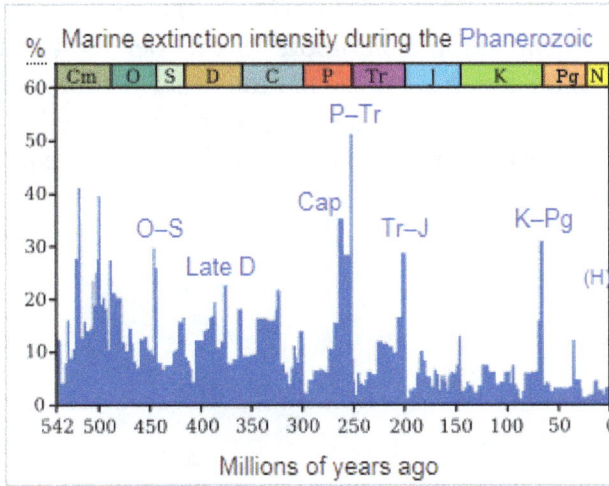

The blue graph shows the apparent *percentage* (not the absolute number) of marine animal genera becoming extinct during any given time interval. It does not represent all marine species, just those that are readily fossilized. The labels of the traditional "Big Five" extinction events and the more recently recognised End-Capitanian extinction event are clickable hyperlinks

Modern Extinctions

According to a 1998 survey of 400 biologists conducted by New York's American Museum of Natural History, nearly 70% believed that the Earth is currently in the early stages of a human-caused mass extinction, known as the Holocene extinction. In that survey, the same proportion of respondents agreed with the prediction that up to 20% of all living populations could become extinct within 30 years (by 2028). A 2014 special edition of *Science* declared there is widespread consensus on the issue of human-driven mass species extinctions.

Biologist E. O. Wilson estimated in 2002 that if current rates of human destruction of the biosphere continue, one-half of all plant and animal species of life on earth will be extinct in 100 years. More significantly, the current rate of global species extinctions is estimated as 100 to 1000 times "background" rates (the average extinction rates in the evolutionary time scale of planet Earth), while future rates are likely 10,000 times higher. However, some groups are going extinct much faster. Biologists Paul R. Ehrlich and Stuart Pimm contend that human population growth is one of the main drivers of the modern extinction crisis.

History of Scientific Understanding

For much of history, the modern understanding of extinction as the end of a species was incompatible with the prevailing worldview. Through the 18th century, much of Western society adhered to the belief that the world was created by God and as such was complete and perfect. This concept reached its heyday in the 1700s with the peak popularity of a theological concept called the Great Chain of Being, in which all life on earth, from the tiniest microorganism to God, is linked in a continuous chain. The extinction of a species was impossible under this model, as it would create gaps or missing links in the chain and destroy the natural order. Thomas Jefferson was a firm supporter of the Great Chain of Being and an opponent of extinction, famously denying the extinction of the wooly mammoth on the grounds that nature never allows a race of animals to become extinct.

Dilophosaurus, one of the many extinct dinosaur genera. The cause of the Cretaceous–Paleogene extinction event is a subject of much debate amongst researchers

A series of fossils were discovered in the late 17th century that appeared unlike any living species. As a result, the scientific community embarked on a voyage of creative rationalization, seeking to understand what had happened to these species within a framework that did not account for total extinction. In October 1686, Robert Hooke presented an impression of a nautilus to the Royal Society that was more than two feet in diameter, and morphologically distinct from any known living species. Hooke theorized that this was simply because the species lived in the deep ocean and no one had discovered them yet. While he contended that it was possible a species could be "lost", he thought this highly unlikely. Similarly, in 1695, Thomas Molyneux published an account of enormous antlers found in Ireland that did not belong to any extant taxa in that area. Molyneux reasoned that they came from the North American moose and that the animal had once been common on the British Isles. Rather than suggest that this indicated the possibility of species going extinct, he argued that although organisms could become locally extinct, they could never be entirely lost and would continue to exist in some unknown region of the globe. Using the antlers as evidence for this position, Molyneux described how moose had continued to exist in North America even as they were lost to the British Isles. The antlers were later confirmed to be from the extinct Irish elk *Megaloceros*. Hooke and Molyneux's line of thinking was difficult to disprove. When parts of the world had not been thoroughly examined and charted, scientists could not rule out that animals found only in the fossil record were not simply "hiding" in unexplored regions of the Earth.

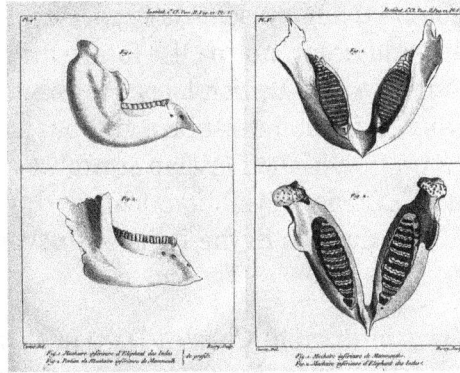

Georges Cuvier compared fossil mammoth jaws to those of living elephants, concluding that they were distinct from any known living species.

Georges Cuvier is credited with establishing the modern conception of extinction in a 1796 lecture to the French Institute, though he would spent most of his career trying to convince the wider scientific community of his theory. Cuvier was a well-regarded geologist, lauded for his ability to reconstruct the anatomy of an unknown species from a few fragments of bone. His primary evidence for extinction came from mammoth skulls found in the Paris basin. Cuvier recognized them as distinct from any known living species of elephant, and argued that it was highly unlikely such an enormous animal would go undiscovered. In 1812, Cuvier, along with Alexandre Bronigniart & Geoffroy Saint-Hilaire, mapped the strata of the Paris basin. They saw alternating saltwater and freshwater deposits, as well as patterns of the appearance and disappearance of fossils throughout the record. From these patterns, Cuvier inferred historic cycles of catastrophic flooding, extinction, and repopulation of the earth with new species.

Cuvier's fossil evidence showed that very different life forms existed in the past than those that exist today, a fact that was accepted by most scientists. The primary debate focused whether this turnover caused by extinction was gradual or abrupt in nature. Cuvier understood extinction to be the result of cataclysmic events that wipe out huge numbers of species, as opposed to the gradual decline of a species over time. His catastrophic view of the nature of extinction garnered him many opponents in the newly emerging school of uniformitarianism.

Jean-Baptist Lamarck, a gradualist and colleague of Cuvier, saw the fossils of different life forms as evidence of the mutable character of species. While Lamarck did not deny the possibility of extinction, he believed that it was exceptional and rare and that most of the change in species over time was due to gradual change. Unlike Cuvier, Lamarck was skeptical that catastrophic events of a scale large enough to cause total extinction were possible. In his geological history of the earth titled Hydrogeologie, Lamarck instead argued that the surface of the earth was shaped by gradual erosion and deposition by water, and that species changed over time in response to the changing environment.

Charles Lyell, a noted geologist and founder of uniformitarianism, believed that past processes should be understood using present day processes. Like Lamarck, Lyell acknowledged that extinction could occur, noting the total extinction of the dodo and the extirpation of indigenous horses to the British Isles. He similarly argued against mass extinctions, believing that any extinction must be a gradual process. Lyell also showed that Cuvier's original interpretation of the Parisian strata was incorrect. Instead of the catastrophic floods inferred by Cuvier, Lyell demonstrated that patterns of saltwater and freshwater deposits, like those seen in the Paris basin, could be formed by a slow rise and fall of sea levels.

The concept of extinction was integral to Charles Darwin's *On the Origin of Species*, with less fit lineages disappearing over time. For Darwin, extinction was a constant side effect of competition. Because of the wide reach of *On the Origin of Species*, it was widely accepted that extinction occurred gradually and evenly (a concept we now refer to as background extinction). It was not until 1982, when David Raup and Jack Sepkoski published their seminal paper on mass extinctions, that Cuvier was vindicated and catastrophic extinction was accepted as an important mechanism. The current understanding of extinction is a synthesis of the cataclysmic extinction events proposed by Cuvier, and the background extinction events proposed by Lyell and Darwin.

Human Attitudes and Interests

Extinction is an important research topic in the field of zoology, and biology in general, and has also become an area of concern outside the scientific community. A number of organizations, such as the Worldwide Fund for Nature, have been created with the goal of preserving species from extinction. Governments have attempted, through enacting laws, to avoid habitat destruction, agricultural over-harvesting, and pollution. While many human-caused extinctions have been accidental, humans have also engaged in the deliberate destruction of some species, such as dangerous viruses, and the total destruction of other problematic species has been suggested. Other species were deliberately driven to extinction, or nearly so, due to poaching or because they were "undesirable", or to push for other human agendas. One example was the near extinction of the American bison, which was nearly wiped out by mass hunts sanctioned by the United States government, to force the removal of Native Americans, many of whom relied on the bison for food.

Biologist Bruce Walsh of the University of Arizona states three reasons for scientific interest in the preservation of species; genetic resources, ecosystem stability, and ethics; and today the scientific community "stress[es] the importance" of maintaining biodiversity.

In modern times, commercial and industrial interests often have to contend with the effects of production on plant and animal life. However, some technologies with minimal, or no, proven harmful effects on *Homo sapiens* can be devastating to wildlife (for example, DDT). Biogeographer Jared Diamond notes that while big business may label environmental concerns as "exaggerated", and often cause "devastating damage", some corporations find it in their interest to adopt good conservation practices, and even engage in preservation efforts that surpass those taken by national parks.

Governments sometimes see the loss of native species as a loss to ecotourism, and can enact laws with severe punishment against the trade in native species in an effort to prevent extinction in the wild. Nature preserves are created by governments as a means to provide continuing habitats to species crowded by human expansion. The 1992 Convention on Biological Diversity has resulted in international Biodiversity Action Plan programmes, which attempt to provide comprehensive guidelines for government biodiversity conservation. Advocacy groups, such as The Wildlands Project and the Alliance for Zero Extinctions, work to educate the public and pressure governments into action.

People who live close to nature can be dependent on the survival of all the species in their environment, leaving them highly exposed to extinction risks. However, people prioritize day-to-day survival over species conservation; with human overpopulation in tropical developing countries, there has been enormous pressure on forests due to subsistence agriculture, including slash-and-burn agricultural techniques that can reduce endangered species's habitats.

Planned Extinction

Completed

- The smallpox virus is now extinct in the wild, although samples are retained in laboratory settings.

- The rinderpest virus, which infected domestic cattle, is now extinct in the wild.

Proposed

The poliovirus is now confined to small parts of the world due to extermination efforts.

Dracunculus medinensis, a parasitic worm which causes the disease dracunculiasis, is now close to eradication thanks to efforts led by the Carter Center.

Treponema pallidum pertenue, a bacterium which causes the disease yaws, is in the process of being eradicated.

Biologist Olivia Judson has advocated the deliberate extinction of certain disease-carrying mosquito species. In a September 25, 2003 *New York Times* article, she advocated "specicide" of thirty mosquito species by introducing a genetic element which can insert itself into another crucial gene, to create recessive "knockout genes". She says that the *Anopheles* mosquitoes (which spread malaria) and *Aedes* mosquitoes (which spread dengue fever, yellow fever, elephantiasis, and other diseases) represent only 30 species; eradicating these would save at least one million human lives per annum, at a cost of reducing the genetic diversity of the family Culicidae by only 1%. She further argues that since species become extinct "all the time" the disappearance of a few more will not destroy the ecosystem: "We're not left with a wasteland every time a species vanishes. Removing one species sometimes causes shifts in the populations of other species—but different need not mean worse." In addition, anti-malarial and mosquito control programs offer little realistic hope to the 300 million people in developing nations who will be infected with acute illnesses this year. Although trials are ongoing, she writes that if they fail: "We should consider the ultimate swatting."

Biologist E. O. Wilson has advocated the eradication of several species of mosquito, including malaria vector Anopheles gambiae. Wilson stated, "I'm talking about a very small number of species that have co-evolved with us and are preying on humans, so it would certainly be acceptable to remove them. I believe it's just common sense."

Cloning

Some, such as Harvard geneticist George M. Church, believe that ongoing technological advances will let us "bring back to life" an extinct species by cloning, using DNA from the remains of that species. Proposed targets for cloning include the mammoth, the thylacine, and the Pyrenean ibex. For this to succeed, enough individuals would have to be cloned, from the DNA of different individuals (in the case of sexually reproducing organisms) to create a viable population. Though bioethical and philosophical objections have been raised, the cloning of extinct creatures seems theoretically possible.

In 2003, scientists tried to clone the extinct Pyrenean ibex (*C. p. pyrenaica*). This attempt failed: of the 285 embryos reconstructed, 54 were transferred to 12 mountain goats and mountain goat-domestic goat hybrids, but only two survived the initial two months of gestation before they too died. In 2009, a second attempt was made to clone the Pyrenean ibex: one clone was born alive, but died seven minutes later, due to physical defects in the lungs.

Biological Dispersal

Biological dispersal refers to both the movement of individuals (animals, plants, fungi, bacteria, etc.) from their birth site to their breeding site ('natal dispersal'), as well as the movement from one breeding site to another ('breeding dispersal'). Dispersal is also used to describe the movement of propagules such as seeds and spores. Technically, dispersal is defined as any movement that has the potential to lead to gene flow. The act of dispersal involves three phases: departure, transfer, settlement and there are different fitness costs and benefits associated with each of these phases. Through simply moving from one habitat patch to another, the dispersal of an individual has consequences not only for individual fitness, but also for population dynamics, population genetics, and species distribution. Understanding dispersal and the consequences both for evolutionary strategies at a species level, and for processes at an ecosystem level, requires understanding on the type of dispersal, the dispersal range of a given species, and the dispersal mechanisms involved.

Wind dispersal of dandelion seeds.

Biological dispersal may be contrasted with geodispersal, which is the mixing of previously isolated populations (or whole biotas) following the erosion of geographic barriers to dispersal or gene flow (Lieberman, 2005; Albert and Reis, 2011).

Dispersal can be distinguished from animal migration (typically round-trip seasonal movement), although within the population genetics literature, the terms 'migration' and 'dispersal' are often used interchangeably.

Types of Dispersal

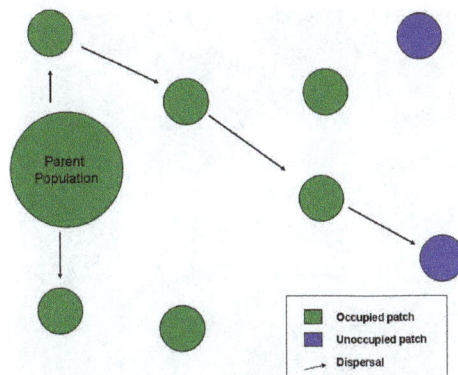

Dispersal from parent population

Some organisms are motile throughout their lives, but others are adapted to move or be moved at precise, limited phases of their life cycles. This is commonly called the dispersive phase of the life cycle. The strategies of organisms' entire life cycles often are predicated on the nature and circumstances of their dispersive phases.

In general there are two basic types of dispersal:

Density-independent dispersal

Organisms have evolved adaptations for dispersal that take advantage of various forms of kinetic energy occurring naturally in the environment. This is referred to as density independent or passive dispersal and operates on many groups of organisms (some invertebrates, fish, insects and sessile organisms such as plants) that depend on animal vectors, wind, gravity or current for dispersal.

Density-dependent dispersal

Density dependent or active dispersal for many animals largely depends on factors such as local population size, resource competition, habitat quality, and habitat size.

Due to population density, dispersal may relieve pressure for resources in an ecosystem, and competition for these resources may be a selection factor for dispersal mechanisms.

Dispersal of organisms is a critical process for understanding both geographic isolation in evolution through gene flow and the broad patterns of current geographic distributions (biogeography).

A distinction is often made between natal dispersal where an individual (often a juvenile) moves away from the place it was born, and breeding dispersal where an individual (often an adult) moves away from one breeding location to breed elsewhere.

Costs and Benefits

Epilobium hirsutum - Seed head

In the broadest sense, dispersal occurs when the fitness benefits of moving outweigh the costs.

There are a number of benefits to dispersal such as locating new resources, escaping unfavorable conditions, avoiding competing with siblings, and avoiding breeding with closely related individuals which could lead to inbreeding depression.

There are also a number of costs associated with dispersal, which can be thought of in terms of four main currencies: energy, risk, time and opportunity. Energetic costs include the extra energy required to move as well as energetic investment in movement machinery (e.g. wings). Risks include increased injury and mortality during dispersal and the possibility of settling in an unfavorable environment. Time spent dispersing is time that often cannot be spent on other activities such as growth and reproduction. Finally dispersal can also lead to outbreeding depression if an individual is better adapted to its natal environment than the one it ends up in. In social animals (such as many birds and mammals) a dispersing individual must find and join a new group, which can lead to loss of social rank.

Dispersal Range

"Dispersal range" refers to the distance a species can move from an existing population or the parent organism. An ecosystem depends critically on the ability of individuals and populations to disperse from one habitat patch to another. Therefore, biological dispersal is critical to the stability of ecosystems.

Environmental Constraints

Few species are ever evenly or randomly distributed within or across landscapes. In general, species significantly vary across the landscape in association with environmental features that influence their reproductive success and population persistence. Spatial patterns in environmental features (e.g. resources) permit individuals to escape unfavorable conditions and seek out new locations. This allows the organism to "test" new environments for their suitability, provided they are within animal's geographic range. In addition, the ability of a species to disperse over a gradually changing environment could enable a population to survive extreme conditions. (i.e. climate change).

As the climate changes, prey and predators have to adapt to survive. This poses a problem for many animals, for example the Southern Rockhopper Penguins. These penguins are able to live and thrive in a variety of climates due to the penguins' phenotypic plasticity. However, they are predicted to respond by dispersal, not adaptation this time. This is explained due to their long life spans and slow microevolution. Penguins in the subantarctic have very different foraging behavior than the subtropical waters, it would be very hard to survive and keep up with the fast changing climate because these behaviors took years to shape.

Dispersal Barriers

A dispersal barrier may mean that the dispersal range of a species is much smaller than the species distribution. An artificial example is habitat fragmentation due to human land use. Natural barriers to dispersal that limit species distribution include mountain ranges and rivers. An example is the separation of the ranges of the two species of chimpanzee by the Congo River.

On the other hand, human activities may also expand the dispersal range of a species by providing new dispersal methods (e.g., ships). Many of them become invasive, like rats and stinkbugs, but some species also have a slightly positive effect to human settlers like honeybees and earthworms.

Dispersal Mechanisms

Most animals are capable of locomotion and the basic mechanism of dispersal is movement from one place to another. Locomotion allows the organism to "test" new environments for their suitability, provided they are within the animal's range. Movements are usually guided by inherited behaviors.

The formation of barriers to dispersal or gene flow between adjacent areas can isolate populations on either side of the emerging divide. The geographic separation and subsequent genetic isolation of portions of an ancestral population can result in speciation.

Plant Dispersal Mechanisms

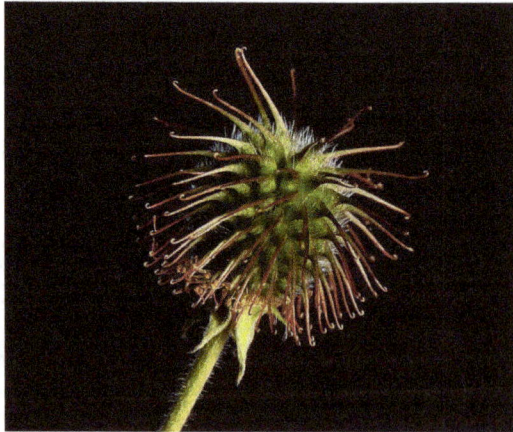

Burs are an example of a seed dispersion mechanism which uses a biotic vector, in this case animals with fur.

Seed dispersal is the movement or transport of seeds away from the parent plant. Plants have limited mobility and consequently rely upon a variety of dispersal vectors to transport their propagules, including both abiotic and biotic vectors. Seeds can be dispersed away from the parent plant individually or collectively, as well as dispersed in both space and time. The patterns of seed dispersal are determined in large part by the dispersal mechanism and this has important implications for the demographic and genetic structure of plant populations, as well as migration patterns and species interactions. There are five main modes of seed dispersal: gravity, wind, ballistic, water and by animals.

Animal Dispersal Mechanisms

Non-motile Animals

There are numerous animal forms that are non—motile, such as sponges, bryozoans, tunicates, sea anemones, corals, and oysters. In common, they are all either marine or aquatic. It may seem curious that plants have been so successful at stationary life on land, while animals have not, but the answer lies in the food supply. Plants produce their own food from sunlight and carbon dioxide—both generally more abundant on land than in water. Animals fixed in place must rely on the surrounding medium to bring food at least close enough to grab, and this occurs in the three-dimensional water environment, but with much less abundance in the atmosphere.

All of the marine and aquatic invertebrates whose lives are spent fixed to the bottom (more or less; anemones are capable of getting up and moving to a new location if conditions warrant) produce dispersal units. These may be specialized "buds", or motile sexual reproduction products, or even a sort of alteration of generations as in certain cnidaria.

Corals provide a good example of how sedentary species achieve dispersion. Corals reproduce by releasing sperm and eggs directly into the water. These release events are coordinated by lunar phase in certain warm months, such that all corals of one or many species on a given reef will release on the same single or several consecutive nights. The released eggs are fertilized, and the resulting zygote develops quickly into a multicellular *planula*. This motile stage then attempts to find a suitable substratum for settlement. Most are unsuccessful and die or are fed upon by zooplankton and bottom dwelling predators such as anemones and other corals. However, untold millions are produced, and a few do succeed in locating spots of bare limestone, where they settle and transform by growth into a polyp. All things being favorable, the single polyp grows into a coral head by budding off new polyps to form a colony.

Motile Animals

The majority of all animals are motile. Although motile animals can, in theory, disperse themselves by their spontaneous and independent locomotive powers, a great many species utilize the existing kinetic energies in the environment, resulting in passive movement. Dispersal by water currents is especially associated with the physically small inhabitants of marine waters known as zooplankton. The term plankton comes from the Greek word meaning "wanderer" or "drifter".

Dispersal by Dormant Stages

Many animal species, especially freshwater invertebrates, are able to disperse by wind or by transfer with an aid of larger animals (birds, mammals or fishes) as dormant eggs, dormant embryos or, in some cases, dormant adult stages. Tardigrades, some rotifers and some copepods are able to withstand desiccation as adult dormant stages. Many other taxa (Cladocera, Bryozoa, Hydra, Copepoda and so on) can disperse as dormant eggs or embryos. Freshwater sponges usually have special dormant propagules called gemmulae for such a dispersal. Many kinds of dispersal dormant stages are able to withstand not only desiccation and low and high temperature, but also action of digestive enzymes during their transfer through digestive tracts of birds and other animals, high concentration of salts and many kinds of toxicants. Such dormant-resistant stages made possible the long-distance dispersal from one water body to another and broad distribution ranges of many freshwater animals.

Quantifying Dispersal

Dispersal is most commonly quantified either in terms of rate or distance.

Dispersal rate (also called migration rate in the population genetics literature) or probability describes the probability than any individual leaves an area or, equivalently, the expected proportion of individual to leave an area.

The dispersal distance is usually described by a dispersal kernel which gives the probability distribution of the distance traveled by any individual. A number of different functions are used for dispersal kernels in theoretical models of dispersal including the negative exponential distribution, extended negative exponential distribution, normal distribution, exponential power distribution, inverse power distribution, and the two-sided power distribution. The inverse power distribution and distributions with 'fat tails' representing long-distance dispersal events (called leptokurtic distributions) are though to best match empirical dispersal data.

Consequences of Dispersal

Dispersal not only has costs and benefits to the dispersing individual but it also has consequences at the level of the population and species as well.

Most populations have a patchy spatial distribution. Dispersal, by moving individuals between different sub-populations, can increase the overall connectivity of the population, helping to minimize the risk of stochastic extinction, since if a sub-population goes extinct by chance, it is likely to be recolonized if the dispersal rate is high. Increased connectivity can also decrease the degree of local adaptation.

Gene Flow

In population genetics, gene flow (also known as gene migration) is the transfer of genetic variation from one population to another. If the rate of gene flow is high enough, then two populations are considered to have equivalent genetic diversity and therefore effectively a single population. It has been shown that it takes only "One migrant per generation" to prevent population diverging due to drift. Gene flow is an important mechanism for transferring genetic diversity among populations. Migrants into or out of a population may result in a change in allele frequencies (the proportion of members carrying a particular variant of a gene), changing the distribution of genetic diversity within the populations. Immigration may also result in the addition of new genetic variants to the established gene pool of a particular species or population. High rates of gene flow can reduce the genetic differentiation between the two groups, increasing homogeneity. For this reason,gene flow has been thought to constrain speciation by combining the gene pools of the groups, and thus, preventing the development of differences in genetic variation that would have led to full speciation.

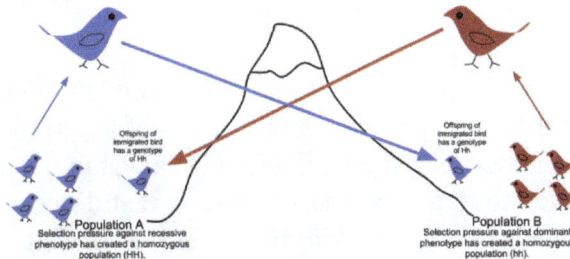

Gene flow is the transfer of alleles from one population to another population through immigration of individuals.

There are a number of factors that affect the rate of gene flow between different populations. Gene flow is expected to be lower in species that have low dispersal or mobility, occur in fragmented habitats, there is long distant between populations, and smaller populations sizes. Mobility plays an important role in the migration rate as a highly mobile individuals tend to have greater migratory potential. Animals tend to be more mobile than plants, although pollen and seeds may be carried great distances by animals or wind. As dispersal distance decreases, gene flow is impeded and inbreeding, measured by the inbreeding coefficient (F), increases.For example, many island populations have low rates of gene flow due to geographically isolated and small population size. The Black Footed Rock Wallaby has several inbred populations that live on various islands off the coast of Australia. The population is so strongly isolated that gene flow is not a possibility leading to high occurrences of inbreeding.

Measuring Gene Flow

Decrease in population size leads to increased divergence due to drift, while migration reduces divergence and inbreeding. Gene flow can be measured by using the effective population size (N_e) and the net migration rate per generation (m). Using the approximation based on the Island model, the effect of migration can be calculated for a population in terms of the degree of genetic differentiation(Fst). This formula accounts fro the proportion of total molecular marker variation among populations, averaged over loci. When there is one migrant per generation, the inbreeding coefficient (Fst) equals 0.2. However, when there is less than 1 migrant per generation (no migration), the inbreeding coefficient rises rapidly resulting in fixation and complete divergence (Fst = 1). The most common Fst is < 0.25. This means there is some migration happening. Measures of population structure range from 0 to 1. When gene flow occurs via migration the deleterious effects of inbreeding can be ameliorated.

$$Fst = 1/(4N_e m + 1)$$

The formula can be modified to solve for the migration rate when Fst is known: $Nm = 1(1/Fst - 1)/4$, Nm = number of migrants .

Barriers to Gene Flow

Allopatric Speciation

When gene flow is blocked by physical barriers, this results in Allopatric speciation or a geographical isolation that does not allow populations of the same species to exchange genetic material. Physical barriers to gene flow are usually, but not always, natural. They may include impassable mountain ranges, oceans, or vast deserts. In some cases, they can be artificial, man-made barriers, such as the Great Wall of China, which has hindered the gene flow of native plant populations. One of these native plants, *Ulmus pumila*, demonstrated a lower prevalence of genetic differentiation than the plants *Vitex negundo, Ziziphus jujuba, Heteropappus hispidus,* and *Prunus armeniaca* whose habitat is located on the opposite side of the Great Wall of China where *Ulmus pumila* grows. This is because *Ulmus pumila* has wind-pollination as its primary means of propagation and the latter-plants carry out pollination through insects. Samples of the same species which grow on either side have been shown to have developed genetic differences, because there is little to no gene flow to provide recombination of the gene pools.

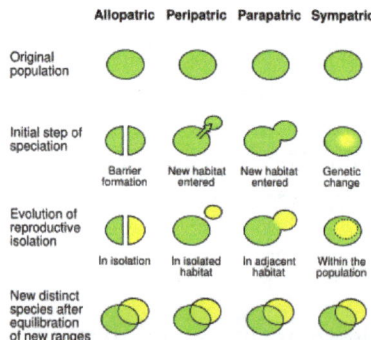

Examples of speciation affecting gene flow.

Sympatric Speciation

Barriers to gene flow need not always be physical. Sympatric speciation happens when new species from the same ancestral species arise along the same range. This is often a result of a reproductive barrier. For example, two palm species of *Howea* found on Lord Howe Island were found to have substantially different flowering times correlated with soil preference, resulting in a reproductive barrier inhibiting gene flow. Species can live in the same environment, yet show very limited gene flow due to reproductive barriers, fragmentation, specialist pollinators, or limited hybridization or hybridization yielding unfit hybrids. A cryptic species is a species that humans cannot tell is different without the use of genetics. Moreover, gene flow between hybrid and wild populations can result in loss of genetic diversity via genetic pollution, assortative mating and out-breeding.

Gene Flow between Species

Horizontal Gene Transfer

Horizontal gene transfer is the transfer of genes between organisms, either through hybridization, antigenic shift, or reassortment is sometimes an important source of genetic variation. Viruses can transfer genes between species. Bacteria can incorporate genes from other dead bacteria, exchange genes with living bacteria, and can exchange plasmids across species boundaries. "Sequence comparisons suggest recent horizontal transfer of many genes among diverse species including across the boundaries of phylogenetic 'domains'. Thus determining the phylogenetic history of a species can not be done conclusively by determining evolutionary trees for single genes."

Biologist Gogarten suggests "the original metaphor of a tree no longer fits the data from recent genome research". Biologists [should] instead use the metaphor of a mosaic to describe the different histories combined in individual genomes and use the metaphor of an intertwined net to visualize the rich exchange and cooperative effects of horizontal gene transfer.

"Using single genes as phylogenetic markers, it is difficult to trace organismal phylogeny in the presence of HGT. Combining the simple coalescence model of cladogenesis with rare HGT events suggest there was no single last common ancestor that contained all of the genes ancestral to those shared among the three domains of life. Each contemporary molecule has its own history and traces back to an individual molecule cenancestor. However, these molecular ancestors were likely to be present in different organisms at different times."

Genetic Pollution

Naturally-evolved, region-specific species can be threatened with extinction through genetic pollution, potentially causing uncontrolled hybridization, introgression and genetic swamping. These processes can lead to homogenization or replacement of local genotypes as a result of either a numerical and/or fitness advantage of introduced plant or animal. Nonnative species can threaten native plants and animals with extinction by hybridization and introgression either through purposeful introduction by humans or through habitat modification, bringing previously isolated species into contact. These phenomena can be especially detrimental for rare species coming into contact with more abundant ones which can occur between island and mainland species. Interbreeding between the species can cause a 'swamping' of the rarer species' gene pool, creating hybrids that supplant the native stock. The extent of this phenomenon is not always apparent from outward appearance alone. While some degree of gene flow occurs in the course of normal evolution, hybridization with or without introgression may threaten a rare species' existence. For example, the Mallard is an abundant species of duck that interbreeds readily with a wide range of other ducks and poses a threat to the integrity of some species.

Examples

Marine iguana of the Galapagos Islands evolved via allopatric speciation, through limited gene flow and geographic isolation.

- Fragmented Population: fragmented landscapes such as the Galapagos Islands are a ideal place for adaptive radiation to occur as a result of differing geography. Darwin's Finches likely experienced allopatric speciation in some part due to differing geography, but that doesn't explain why we see some many different kinds of finches on the same island. This is due to adaptive radiation, or the evolution or varying traits in light of competition for resources. Gene flow moves in the direction of what resources are abundant at a given time.

- Island Population: The Marine Iguana is en endemic species of the Galapagos Islands, but it evolved from a mainland ancestor of land iguana. Due to geographic isolation gene flow between the two species was limited and differing environments caused the Marine Iguana to evolve in order to adapt to the island environment. For instance, they are the only iguana that has evolved the ability to swim.

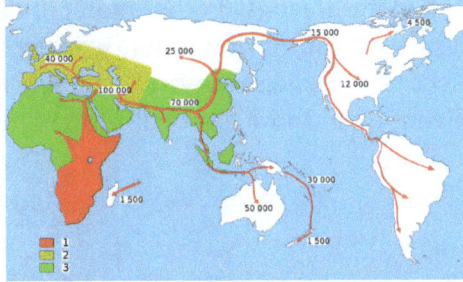

Theorized historic radiation of the first humans throughout the world and various species of homoinids that may have contributed to the modern day humans.

- **Human Populations:** Two theories exist for the human evolution throughout the world. The first is known as the multiregional model in which modern human variation is seen as a product of radiation of *Homo erectus* out of Africa after which local differentiation led to the establishment of regional population as we see them now. Gene flow plays an important role in maintaining a grade of similarities and preventing speciation. In contrast the single origin theory assumes that there was a common ancestral population originating in Africa of *Homo sapiens* which already displayed the anatomical characteristics we see today. This theory minimizes the amount of parallel evolution that is needed.

- **Butterflies:** Comparisons between sympatric and allopatric populations of *Heliconius melpomene*, *H. cydno*, and *H. timareta* revealed a genome-wide trend of increased shared variation in sympatry, indicative of pervasive interspecific gene flow.

- **Plants:** Two species of Monkeyflowers, *mimulus lewsii* and *mimulus cardinalis,* were found to have highly specialized pollinators that acted on major genes resulting in a contribution to the floral evolution and reproductive isolation of these two species. The specialized pollination limited gene flow between the two species, eventually resulting in two different species.

- **Human-mediate gene flow:** The captive genetic management of threatened species is one way in which humans attempt to induce gene flow in ex situ situation. One example is the Giant Panda which is part of a international breeding program in which genetic materials is shared between zoological organizations in order to increase genetic diversity in the small populations. As a result of low reproductive success, artificial insemination with fresh/frozen-thawed sperm was developed which increased cub survival rate. A 2014 study found that high levels of genetic diversity and low levels of inbreeding were estimated in the breeding centers.

Seed Dispersal

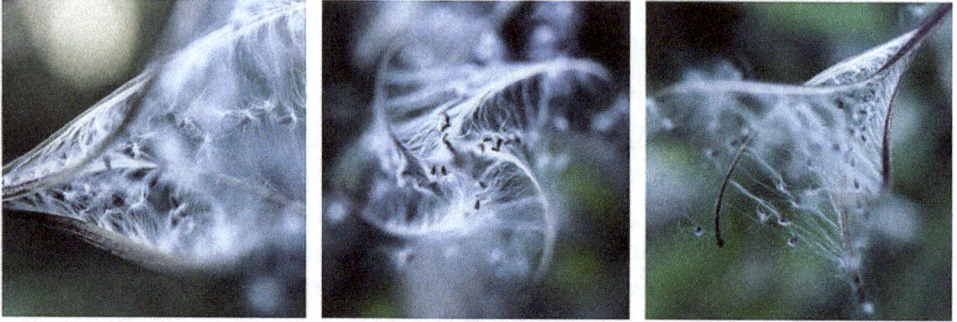

Epilobium hirsutum seed head dispersing seeds

Seed dispersal is the movement or transport of seeds away from the parent plant. Plants have very limited mobility and consequently rely upon a variety of dispersal vectors to transport their propagules, including both abiotic and biotic vectors. Seeds can be dispersed away from the parent plant individually or collectively, as well as dispersed in both space and time. The patterns of seed dispersal are determined in large part by the dispersal mechanism and this has important implications for the demographic and genetic structure of plant populations, as well as migration patterns and species interactions. There are five main modes of seed dispersal: gravity, wind, ballistic, water, and by animals. Some plants are serotinous and only disperse their seeds in response to an environmental stimulus.

Benefits

Seed dispersal is likely to have several benefits for plant species. First, seed survival is often higher away from the parent plant. This higher survival may result from the actions of density-dependent seed and seedling predators and pathogens, which often target the high concentrations of seeds beneath adults. Competition with adult plants may also be lower when seeds are transported away from their parent.

Seed dispersal also allows plants to reach specific habitats that are favorable for survival, a hypothesis known as directed dispersal. For example, *Ocotea endresiana* (Lauraceae) is a tree species from Latin America which is dispersed by several species of birds, including the three-wattled bellbird. Male bellbirds perch on dead trees in order to attract mates, and often defecate seeds beneath these perches where the seeds have a high chance of survival because of high light conditions and escape from fungal pathogens. In the case of fleshy-fruited plants, seed-dispersal in animal guts (endozoochory) often enhances the amount, the speed, and the asynchrony of germination, which can have important plant benefits.

Seeds dispersed by ants (myrmecochory) are not only dispersed short distances but are also buried underground by the ants. These seeds can thus avoid adverse environmental effects such as fire or drought, reach nutrient-rich microsites and survive longer than other seeds. These features are peculiar to myrmecochory, which may thus provide additional benefits not present in other dispersal modes.

Finally, at another scale, seed dispersal may allow plants to colonize vacant habitats and even new geographic regions. Dispersal distances and deposition sites depend on the movement range of the disperser, and longer dispersal distances are sometimes accomplished through diplochory, the sequential dispersal by two or more different dispersal mechanisms. In fact, recent evidence suggests that the majority of seed dispersal events involves more than one dispersal phase.

Types

Seed dispersal is sometimes split into *autochory* (when dispersal is attained using the plant's own means) and *allochory* (when obtained through external means).

Long Distance Dispersal of Seeds

Long distance seed dispersal is a type of spatial dispersal that is currently defined by two forms, proportional and actual distance. A plants fitness and survival may heavily depend on this method of seed dispersal depending on certain environmental factors. The first form of LDD, proportional distance, measures the percentage of seeds (1% out of total number of seeds produced) that travel the farthest distance out of a 99% probability distribution. The proportional definition of LDD is in actuality a descriptor for more extreme dispersal events. An example of LDD would be that of a plant developing a specific dispersal vector or morphology in order to allow for the dispersal of its seeds over a great distance. The actual or absolute method identifies LDD as a literal distance. It classifies 1 km as the threshold distance for seed dispersal. Here, threshold means the minimum distance a plant can disperse its seeds and have it still count as LDD. There is a second, unmeasurable, form of LDD besides proportional and actual. This is known as the non-standard form. Non-standard LDD is when seed dispersal occurs completely at random. An example of this would be if the lemur dependent dispersal of seeds from the deciduous trees of Madagascar were to wash ashore the coastline of South Africa via the attachment of mermaid purses laid by a shark or common skate. A driving factor for the evolutionary significance of LDD is that it increases plant fitness by decreasing neighboring plant competition for offspring. However, it is still unclear today as to how specific traits, conditions and trade-offs (particularly within short seed dispersal) effect LDD evolution.

Autochory

Autochorous plants disperse their seed without any help from an external vector, as a result this limits plants considerably as to the distance they can disperse their seed. Two other types of autochory not described in detail here are blastochory, where the stem of the plant crawls along the ground to deposit its seed far from the base of the plant, and herpochory (the seed crawls by means of trichomes and changes in humidity).

Gravity

Barochory or the plant use of gravity for dispersal is a simple means of achieving seed dispersal. The effect of gravity on heavier fruits causes them to fall from the plant when ripe. Fruits exhibiting this type of dispersal include apples, coconuts and passionfruit and those with harder shells (which often roll away from the plant to gain more distance). Gravity dispersal also allows for later transmission by water or animal.

Ballistic Dispersal

Ballochory is a type of dispersal where the seed is forcefully ejected by explosive dehiscence of the fruit. Often the force that generates the explosion results from turgor pressure within the fruit or due to internal tensions within the fruit. Some examples of plants which disperse their seeds autochorously include: *Impatiens spp., Arceuthobium spp., Ecballium spp., Geranium spp., Cardamine hirsuta* and others. An exceptional example of ballochory is *Hura crepitans*—this plant is commonly called the dynamite tree due to the sound of the fruit exploding. The explosions are powerful enough to throw the seed up to 100 meters.

Allochory

Allochory refers to any of many types of seed dispersal where a vector or secondary agent is used to disperse seeds. This vectors may include wind, water, animals or others.

Wind Dispersal

Wind dispersal of dandelion seeds

Entada phaseoloides – Hydrochory

Wind dispersal (*anemochory*) is one of the more primitive means of dispersal. Wind dispersal can take on one of two primary forms: seeds can float on the breeze or alternatively, they can flutter to the ground. The classic examples of these dispersal mechanisms, in the temperate northern hemisphere, include dandelions, which have a feathery pappus attached to their seeds and can be dispersed long distances, and maples, which have winged seeds (samaras) and flutter to the ground. An important constraint on wind dispersal is the need for abundant seed production to maximize the likelihood of a seed landing in a site suitable for germination. There are also strong evolutionary constraints on this dispersal mechanism. For instance, Cody and Overton (1996) found that species in the Asteraceae on islands tended to have reduced dispersal capabilities (i.e., larger seed mass and smaller pappus) relative to the same species on the mainland. Also, *Helonias bullata*, a species of perennial herb native to the United States, evolved to utilize wind dispersal as the primary seed dispersal mechanism; however, limited wind in its habitat prevents the seeds to successfully disperse away from its parents, resulting in clusters of population. Reliance on wind dispersal is common among many weedy or ruderal species. Unusual mechanisms of wind dispersal include tumbleweeds, where the entire plant is blown by the wind. *Physalis* fruits, when not fully ripe, may sometimes be dispersed by wind due to the space between the fruit and the covering calyx which acts as air bladder.

Water

Many aquatic (water dwelling) and some terrestrial (land dwelling) species use *hydrochory*, or seed dispersal through water. Seeds can travel for extremely long distances, depending on the specific mode of water dispersal; this especially applies to fruits which are waterproof and float.

The water lily is an example of such a plant. Water lilies' flowers make a fruit that floats in the water for a while and then drops down to the bottom to take root on the floor of the pond. The seeds of palm trees can also be dispersed by water. If they grow near oceans, the seeds can be transported by ocean currents over long distances, allowing the seeds to be dispersed as far as other continents.

Mangrove trees grow directly out of the water; when their seeds are ripe they fall from the tree and grow roots as soon as they touch any kind of soil. During low tide, they might fall in soil instead of water and start growing right where they fell. If the water level is high, however, they can be carried far away from where they fell. Mangrove trees often make little islands as dirt and other things collect in their roots, making little bodies of land.

A special review for oceanic waters hydrochory can be seen at oceanic dispersal.

The "bill" and seed dispersal mechanism of *Geranium pratense*

By Animals

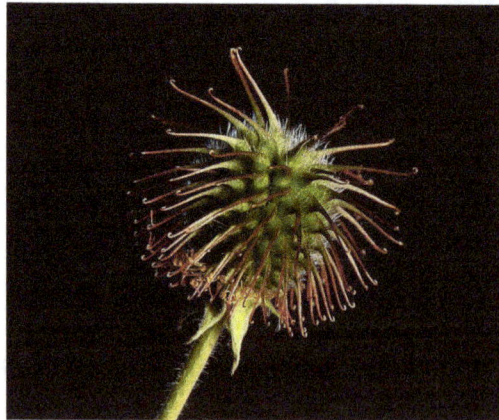

The small hooks on the surface of a bur enable attachment to animal fur for dispersion.

Animals can disperse plant seeds in several ways, all named *zoochory*. Seeds can be transported on the outside of vertebrate animals (mostly mammals), a process known as *epizoochory*. Plant species transported externally by animals can have a variety of adaptations for dispersal, including adhesive mucus, and a variety of hooks, spines and barbs. A typical example of an epizoochorous plant is *Trifolium angustifolium*, a species of Old World clover which adheres to animal fur by means of stiff hairs covering the seed. Epizoochorous plants tend to be herbaceous plants, with many representative species in the families Apiaceae and Asteraceae. However, epizoochory is a relatively rare dispersal syndrome for plants as a whole; the percentage of plant species with seeds adapted for transport on the outside of animals is estimated to be below 5%. Nevertheless, epizoochorous transport can be highly effective if seeds attach to wide-ranging animals. This form of seed dispersal has been implicated in rapid plant migration and the spread of invasive species.

Seed dispersal via ingestion by vertebrate animals (mostly birds and mammals), or *endozoochory*, is the dispersal mechanism for most tree species. Endozoochory is generally a coevolved mutualistic relationship in which a plant surrounds seeds with an edible, nutritious fruit as a good food for animals that consume it. Birds and mammals are the most important seed dispersers, but a wide variety of other animals, including turtles and fish, can transport viable seeds. The exact percentage of tree species dispersed by endozoochory varies between habitats, but can range to over 90% in some tropical rainforests. Seed dispersal by animals in tropical rainforests has received much attention, and this interaction is considered an important force shaping the ecology and evolution of vertebrate and tree populations. In the tropics, large animal seed dispersers (such as tapirs, chimpanzees and hornbills) may disperse large seeds with few other seed dispersal agents. The extinction of these large frugivores from poaching and habitat loss may have negative effects on the tree populations that depend on them for seed dispersal. A variation of endozoochory is regurgitation rather than all the way through the digestive track.

Seed dispersal by ants (*myrmecochory*) is a dispersal mechanism of many shrubs of the southern hemisphere or understorey herbs of the northern hemisphere. Seeds of myrmecochorous plants have a lipid-rich attachment called the elaiosome, which attracts ants. Ants carry such seeds into their colonies, feed the elaiosome to their larvae and discard the otherwise intact seed in an underground chamber. Myrmecochory is thus a coevolved mutualistic relationship between plants and seed-disperser ants. Myrmecochory has independently evolved at least 100 times in flowering plants and is estimated to be present in at least 11 000 species, but likely up to 23 000 or 9% of all species of flowering plants. Myrmecochorous plants are most frequent in the fynbos vegetation of the Cape Floristic Region of South Africa, the kwongan vegetation and other dry habitat types of Australia, dry forests and grasslands of the Mediterranean region and northern temperate forests of western Eurasia and eastern North America, where up to 30–40% of understorey herbs are myrmecochorous.

Seed predators, which include many rodents (such as squirrels) and some birds (such as jays) may also disperse seeds by hoarding the seeds in hidden caches. The seeds in caches are usually well-protected from other seed predators and if left uneaten will grow into new plants. In addition, rodents may also disperse seeds via seed spitting due to the presence of secondary metabolites in ripe fruits. Finally, seeds may be secondarily dispersed from seeds deposited by primary animal dispersers, a process known as diplochory. For example, dung beetles are known to disperse seeds from clumps of feces in the process of collecting dung to feed their larvae.

Other types of zoochory are *chiropterochory* (by bats), *malacochory* (by molluscs, mainly terrestrial snails), *ornithochory* (by birds) and *saurochory* (by non-bird sauropsids). Zoochory can occur in more than one phase, for example through *diploendozoochory*, where a primary disperser (an animal that ate a seed) along with the seeds it is carrying is eaten by a predator that then carries the seed further before depositing it.

By Humans

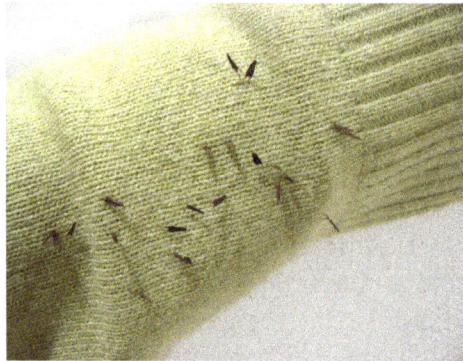

Epizoochory in *Bidens tripartita*; the seeds have attached to the clothes of a human.

Dispersal by humans (*anthropochory*) used to be seen as a form of dispersal by animals. Recent research points out that human dispersers differ from animal dispersers by a much higher mobility based on the technical means of human transport. Dispersal by humans on the one hand may act on large geographical scales and lead to invasive species. On the other hand, dispersal by humans also acts on smaller, regional scales and drives the dynamics of existing biological populations. Humans may disperse seeds by many various means and some surprisingly high distances have been repeatedly measured. Examples are: dispersal on human clothes (up to 250 m), on shoes (up to 5 km) or by cars (regularly ~ 250 m, singles cases > 100 km).

Deliberate seed dispersal also occurs as seed bombing. This has risks as unsuitable provenance may introduce genetically unsuitable plants to new environments.

Consequences

Seed dispersal has many consequences for the ecology and evolution of plants. Dispersal is necessary for species migrations, and in recent times dispersal ability is an important factor in whether or not a species transported to a new habitat by humans will become an invasive species. Dispersal is also predicted to play a major role in the origin and maintenance of species diversity. For example, myrmecochory increased the rate of diversification more than twofold in plant groups in which it has evolved because myrmecochorous lineages contain more than twice as many species as their non-myrmecochorous sister groups. Dispersal of seeds away from the parent organism has a central role in two major theories for how biodiversity is maintained in natural ecosystems, the Janzen-Connell hypothesis and recruitment limitation. Seed dispersal is essential in allowing forest migration of flowering plants.

In addition, the speed and direction of wind are highly influential in the dispersal process and in turn the deposition patterns of floating seeds in the stagnant water bodies. The transportation of seeds is led by the wind direction. This effects colonization situated on the banks of a river or to wetlands adjacent to streams relative to the distinct wind directions. The wind dispersal process can also effect connections between water bodies. Essentially, wind plays a larger role in the dispersal of waterborne seeds in a short period of time, days and seasons, but the ecological process allows the process to become balanced throughout a time period of several years. The time period of which the dispersal occurs is essential when considering the consequences of wind on the ecological process.

Animal Migration

Mexican free-tailed bats on their long aerial migration

Animal migration is the relatively long-distance movement of individuals, usually on a seasonal basis. It is found in all major animal groups, including birds, mammals, fish, reptiles, amphibians, insects, and crustaceans. The trigger for the migration may be local climate, local availability of food, the season of the year or for mating reasons. To be counted as a true migration, and not just a local dispersal or irruption, the movement of the animals should be an annual or seasonal occurrence, such as Northern hemisphere birds migrating south for the winter; wildebeest migrating annually for seasonal grazing; or a major habitat change as part of their life, such as young Atlantic salmon or Sea lamprey leaving the river of their birth when they have reached a few inches in size.

Overview

Migration can take very different forms in different species, and as such there is no simple accepted definition of migration. One of the most commonly used definitions, proposed by Kennedy is Migratory behavior is persistent and straightened out movement effected by the animal's own locomotory exertions or by its active embarkation upon a vehicle. It depends on some temporary inhibition of station keeping responses but promotes their eventual disinhibition and recurrence.

Migration encompasses four related concepts: persistent straight movement; relocation of an individual on a greater scale (both spatially and temporally) than its normal daily activities; seasonal to-and-fro movement of a population between two areas; and movement leading to the redistribution of individuals within a population. Migration can be either obligate, meaning individuals must migrate, or facultative, meaning individuals can "choose" to migrate or not. Within a migratory species or even within a single population, often not all individuals migrate. *Complete migration* is when all individuals migrate, *partial migration* is when some individuals migrate while others do not, and *differential migration* is when the difference between migratory and non-migratory individuals is based on age or sex (for example).

A Christmas Island red crab on its migration

While most migratory movements occur on an annual cycle, some daily movements are also referred to as migration. Many aquatic animals make a Diel vertical migration, travelling a few hundred metres up and down the water column, while some jellyfish make daily horizontal migrations, traveling a few hundred metres across a lake.

Irregular (non-cyclical) migrations such as irruptions can occur under pressure of famine, overpopulation of a locality, or some more obscure influence.

In Specific Groups

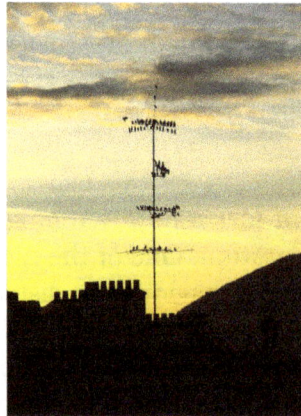

Flocks of birds assembling before migration southwards

Different kinds of animal migrate in different ways.

In Birds

Approximately 1,800 of the world's 10,000 bird species migrate long distances each year in response to the seasons. Many of these migrations are north-south, with species feeding and breeding in high northern latitudes in the summer, and moving some hundreds of kilometres south for the winter. Some species extend this strategy to migrate annually between the Northern and Southern Hemispheres. The Arctic tern is famous for its migration; it flies from its Arctic breeding grounds to the Antarctic and back again each year, a distance of at least 19,000 km (12,000 mi), giving it two summers every year.

In Fish

Many species of salmon migrate up rivers to spawn

Most fish species are relatively limited in their movements, remaining in a single geographical area and making short migrations for wintering, to spawn, or to feed. A few hundred species migrate long distances, in some cases of thousands of kilometres. About 120 species of fish, including several species of salmon, migrate between saltwater and freshwater (they are 'diadromous').

Forage fish such as herring and capelin migrate around substantial parts of the North Atlantic ocean. The capelin for example spawn around the southern and western coasts of Iceland; their larvae drift clockwise around Iceland, while the fish swim northwards towards Jan Mayen island to feed, and return to Iceland parallel with Greenland's east coast.

In the 'sardine run', billions of Southern African pilchard *Sardinops sagax* spawn in the cool waters of the Agulhas Bank and move northward along the east coast of South Africa between May and July.

In Insects

Some winged insects such as locusts and certain butterflies and dragonflies with strong flight migrate long distances. Among the dragonflies, species of *Libellula* and *Sym-*

petrum are known for mass migration, while *Pantala flavescens*, known as the globe skimmer or wandering glider dragonfly, makes the longest ocean crossing of any insect, between India and Africa. Exceptionally, swarms of the desert locust, *Schistocerca gregaria*, flew westwards across the Atlantic Ocean for 4500 km during October 1988, using air currents in the Inter-Tropical Convergence Zone.

An aggregation of migratory *Pantala flavescens* dragonflies, known as globe skimmers, in Coorg, India

In some migratory butterflies, such as the monarch butterfly and the painted lady, no individual completes the whole migration. Instead the butterflies mate and reproduce on the journey, and successive generations travel the next stage of the migration.

In other Animals

Wildebeest on the Serengeti 'great migration'

Mass migration occurs in mammals such as the Serengeti 'great migration', an annual circular pattern of movement with some 1.7 million wildebeest and hundreds of thousands of other large game animals including gazelles and zebra.

Migration is important in other mammals including Cetaceans, whales, dolphins and porpoises. Long-distance migrations occur in some bats, notably the mass migration of the Mexican free-tailed bat between Oregon and southern Mexico.

Some reptiles and amphibians migrate.

Some crustaceans migrate, most spectacularly the Christmas Island red crab which moves en masse each year by the million.

Tracking Migration

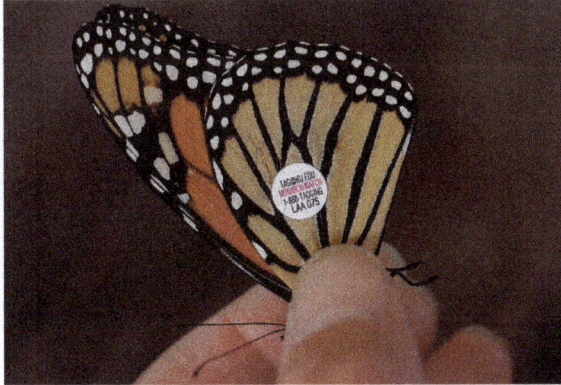

A migratory butterfly, a monarch, tagged for identification

Scientists gather observations of animal migration by tracking their movements. Animals were traditionally tracked with identification tags such as bird rings for later recovery; no information was obtained about the actual route followed between release and recovery, and only a small fraction of tagged individuals were generally recovered. More convenient, therefore, are electronic devices such as radio tracking collars which can be followed by radio, whether handheld, in a vehicle or aircraft, or by satellite. Tags can include a GPS receiver, enabling accurate positions to be broadcast at regular intervals, but these are inevitably heavier and more expensive than devices without GPS. An alternative is the Argos Doppler tag, also called a 'Platform Transmitter Terminal' (PTT) which sends regularly to the polar-orbiting Argos satellites; using Doppler shift, the animal's location can be estimated, relatively roughly compared to GPS, but at lower cost and weight.

Radio tracking tags can be fitted to insects including dragonflies and bees.

In Culture

Before the phenomenon of animal migration was understood, various folklore and erroneous explanations sprang up to account for the disappearance or sudden arrival of birds in an area. In Ancient Greece, Aristotle proposed that robins turned into redstarts when summer arrived. The barnacle goose was explained in European Medieval bestiaries and manuscripts as either growing like fruit on trees, or developing from goose barnacles on pieces of driftwood. Another example is the swallow, which was once thought, even by naturalists such as Gilbert White, to hibernate either underwater, buried in muddy riverbanks, or in hollow trees.

Endemism

Endemism is the ecological state of a species being unique to a defined geographic location, such as an island, nation, country or other defined zone, or habitat type; organisms that are indigenous to a place are not endemic to it if they are also found elsewhere. The extreme opposite of endemism is cosmopolitan distribution. An alternative term for a species that is endemic is precinctive, which applies to species (and subspecific categories) that are restricted to a defined geographical area.

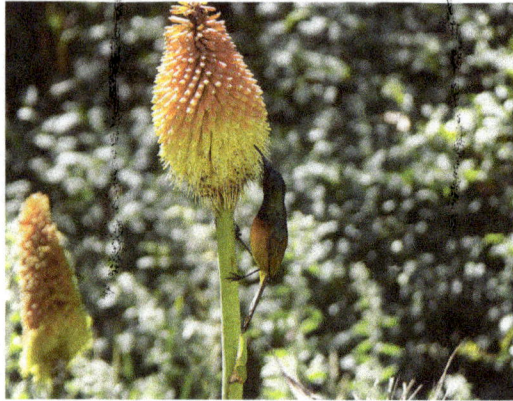

The orange-breasted sunbird (*Nectarinia violacea*) is exclusively found in fynbos vegetation.

The word *endemic* is from New Latin *endēmicus*, from Greek *endēmos*, "native". *Endēmos* is formed of *en* meaning "in", and *dēmos* meaning "the people". The term, *precinctive*, has been suggested by some scientists, and was first used in botany by MacCaughey in 1917. It is the equivalent of "endemism". *Precinction* was perhaps first used by Frank and McCoy. *Precinctive* seems to have been coined by David Sharp when describing the Hawaiian fauna in 1900: "I use the word precinctive in the sense of 'confined to the area under discussion' ... 'precinctive forms' means those forms that are confined to the area specified." That definition excludes artificial confinement of examples by humans in far-off botanical gardens or zoological parks.

Bicolored frog (*Clinotarsus curtipes*) is endemic to the Western Ghats of India

Overview

Physical, climatic, and biological factors can contribute to endemism. The orange-breasted sunbird is exclusively found in the fynbos vegetation zone of southwestern South Africa. The glacier bear is found only in limited places in Southeast Alaska. Political factors can play a part if a species is protected, or actively hunted, in one jurisdiction but not another.

There are two subcategories of endemism: paleoendemism and neoendemism. Paleoendemism refers to species that were formerly widespread but are now restricted to a smaller area. Neoendemism refers to species that have recently arisen, such as through divergence and reproductive isolation or through hybridization and polyploidy in plants.

Endemic types or species are especially likely to develop on geographically and biologically isolated areas such as islands and remote island groups, such as Hawaii, the Galápagos Islands, and Socotra; they can equally develop in biologically isolated areas such as the highlands of Ethiopia, or large bodies of water far from other lakes, like Lake Baikal.

Endemics can easily become endangered or extinct if their restricted habitat changes, particularly—but not only—due to human actions, including the introduction of new organisms. There were millions of both Bermuda petrels and "Bermuda cedars" (actually *junipers*) in Bermuda when it was settled at the start of the seventeenth century. By the end of the century, the petrels were thought extinct. Cedars, already ravaged by centuries of shipbuilding, were driven nearly to extinction in the twentieth century by the introduction of a parasite. Bermuda petrels and cedars are now rare, as are other species endemic to Bermuda.

Threats to Highly Endemistic Regions

Principal causes of habitat degradation and loss in highly endemistic ecosystems include agriculture, urban growth, surface mining, mineral extraction, logging operations and slash-and-burn agriculture.

Cosmopolitan Distribution

Orcinus orca and its range

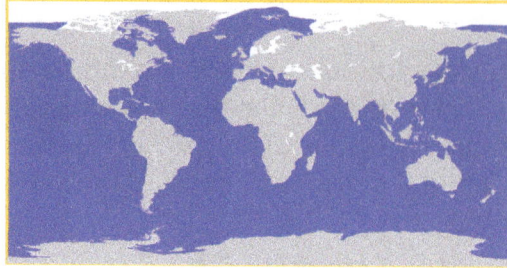

In biogeography, a taxon is said to have a cosmopolitan distribution if its range extends across all or most of the world in appropriate habitats. Such a taxon is said to exhibit cosmopolitanism or cosmopolitism. The opposite extreme is endemism.

Related Terms and Concepts

The term pandemism also is in use, but not all authors are consistent in the sense in which they use the term; some speak of pandemism mainly in referring to diseases and pandemics, and some as a term intermediate between endemism and cosmopolitanism, in effect regarding pandemism as subcosmopolitanism. This means near cosmopolitanism, but with major gaps in the distribution, say, complete absence from Australia. Terminology varies, and there is some debate whether the true opposite of endemism is pandemism or cosmopolitism.

Aspects and Degrees

The term "cosmopolitan distribution" usually should not be taken literally, because it often is applied loosely in various contexts. Commonly the intention is not to include polar regions, extreme altitudes, oceans, deserts, or small, isolated islands. For example, the housefly is nearly as cosmopolitan as any animal species, but it is neither oceanic nor polar in its distribution. Similarly, the term "cosmopolitan weed" implies no more than that the plant in question occurs on all continents except Antarctica; it is not meant to suggest that the species is present in all regions of every continent.

Oceanic and Terrestrial

Another concept in biogeography is that of oceanic cosmopolitanism and endemism. Though there is a temptation to regard the World Ocean as a medium without biological boundaries, this is far from reality; many physical and biological barriers interfere with either the spread or continued residence of many species. For example, temperature gradients prevent free migration of tropical species between the Atlantic and Indian-plus-Pacific oceans, even though there is open passage past continental masses such as the Americas and Africa/Eurasia. Again, as far as many species are concerned, the Southern Ocean and the Northern marine regions are completely isolated from each other by the intolerable temperatures of the tropical regions. In the light of such considerations, it is no surprise to find that endemism and cosmopolitanism are quite as marked in the oceans as on land.

Ecological Delimitation

Another aspect of cosmopolitanism is that of ecological limitations. A species that is apparently cosmopolitan because it occurs in all oceans, might in fact occupy only littoral zones, or only particular ranges of depths, or only estuaries for example. Analogously, terrestrial species might be present only in forests, or mountainous regions, or sandy arid regions or the like. Such distributions might be patchy, or extended, but narrow. Factors of such a nature are taken widely for granted, so they seldom are mentioned explicitly in mentioning cosmopolitan distributions.

Regional and Temporal Variation in Populations

Cosmopolitanism of a particular species or variety should not be confused with cosmopolitanism of higher taxa. For example, the family Myrmeleontidae is cosmopolitan in the sense that every continent except Antarctica is home to some indigenous species within the Myrmeleontidae, but nonetheless no one species, nor even genus, of the Myrmeleontidae is cosmopolitan. Conversely, partly as a result of human introduction of unnatural apiculture to the New World, *Apis mellifera* probably is the only cosmopolitan member of its family; the rest of the family Apidae have modest distributions.

Even where a cosmopolitan population is recognised as a single species, such as indeed *Apis mellifera*, there generally will be variation between regional sub-populations. Such variation commonly is at the level of subspecies, varieties or morphs, whereas some variation is too slight or inconsistent for formal recognition.

For an example of subspecific variation, consider the so-called "African killer bee", which is the subspecies *Apis mellifera scutellata*, and the Cape bee, which is the subspecies *Apis mellifera capensis*; both of them are in the same cosmopolitan species *Apis mellifera*, but their ranges barely overlap.

Other cosmopolitan species, such as the osprey and house sparrow, present similar examples, but in yet other species there are less familiar complications: some migratory birds such as the Arctic tern occur from the Arctic to the Southern Ocean, but at any one season of the year they are likely to be largely in passage or concentrated at only one end of the range. Also, some such species breed only at one end of the range. Seen purely as an aspect of cosmopolitanism, such distributions could be seen as temporal, seasonal variations.

Other complications of cosmopolitanism on a planet too large for local populations to interbreed routinely with each other, lead to genetic effects such as ring species, for example in the *Larus* gulls. They also lead to the formation of clines such as in Drosophila.

Ancient and Modern

Cosmopolitan distributions can be observed both in extinct and extant species. For example, *Lystrosaurus* was cosmopolitan in the Early Triassic after a mass extinction.

In the modern world, the killer whale has a cosmopolitan distribution, extending over most of the Earth's oceans. The wasp *Copidosoma floridanum* is another example, as it is found around the world. Other examples include humans, cats, dogs, orchids, the foliose lichen *Parmelia sulcata*, and the mollusc genus *Mytilus*. The term can also apply to some diseases. It may result from a broad range of environmental tolerances or from rapid dispersal compared to the time needed for evolution.

References

- Newman, Mark (1997). "A model of mass extinction". Journal of Theoretical Biology. 189: 235–252. doi:10.1006/jtbi.1997.0508

- Staff (2 May 2016). "Researchers find that Earth may be home to 1 trillion species". National Science Foundation. Retrieved 6 May 2016

- Gavrilets, S.; Losos, J. B. (2009). "Adaptive radiation: contrasting theory with data". Science. 323 (5915): 732–737. PMID 19197052. doi:10.1126/science.1157966

- Fernández, L. (2015). "Source-sink dynamics shapes the spatial distribution of soil protists in an arid shrubland of northern Chile". Journal of Arid Environment. 113: 121–125. doi:10.1016/j.jaridenv.2014.10.007

- Mills, L. Scott (2009-03-12). Conservation of Wildlife Populations: Demography, Genetics and Management. John Wiley & Sons. p. 13. ISBN 9781444308938

- Quince, C.; et al. "Deleting species from model food webs" (PDF). Archived from the original (PDF) on 2006-09-25. Retrieved 2007-02-15

- Irschick, Duncan J.; et al. (1997). "A comparison of evolutionary radiations in mainland and Caribbean Anolis lizards". Ecology. 78 (7): 2191–2203. doi:10.2307/2265955

- Chapman, D. S.; Dytham, C. & Oxford, G. S. (2007). "Modelling population redistribution in a leaf beetle: an evaluation of alternative dispersal functions". Journal of Animal Ecology. 76 (1): 36–44. PMID 17184351. doi:10.1111/j.1365-2656.2006.01172.x

- Inwood, Stephen (2005-05-03). The Forgotten Genius: The Biography of Robert Hooke, 1635–1703. MacAdam/Cage Publishing. ISBN 9781596921153

- Lee, Anita. "The Pleistocene Overkill Hypothesis Archived October 14, 2006, at the Wayback Machine.." University of California at Berkeley Geography Program.'.' Retrieved January 11, 2007

- Raup, David M.; J. John Sepkoski Jr. (March 1982). "Mass extinctions in the marine fossil record". Science. 215 (4539): 1501–3. Bibcode:1982Sci...215.1501R. PMID 17788674. doi:10.1126/science.215.4539.1501

- Cody, M.L. & Overton, J.M. (1996). "Short-term evolution of reduced dispersal in island plant populations". Journal of Ecology. 84: 53–61. JSTOR 2261699. doi:10.2307/2261699

- Ehrlich, Anne (1981). Extinction: The Causes and Consequences of the Disappearance of Species. Random House, New York. ISBN 0-394-51312-6

- International Programme on Chemical Safety (1989). "DDT and its Derivatives – Environmental Aspects". Environmental Health Criteria 83. Retrieved September 20, 2006

5

Understanding Speciation

Speciation is the process by which the population of the same species evolves into separate species. The types of speciation that occurs are peripatric speciation, sympatric speciation, ecological speciation, etc. This section has been carefully written to provide an easy understanding of the varied facets of speciation.

Speciation

Speciation is the evolutionary process by which biological populations evolve to become distinct species. The biologist Orator F. Cook coined the term 'speciation' in 1906 for the splitting of lineages or "cladogenesis", as opposed to "anagenesis" or "phyletic evolution" within lineages. Charles Darwin was the first to describe the role of natural selection in speciation in his 1859 book *The Origin of Species*. He also identified sexual selection as a likely mechanism, but found it problematic.

There are four geographic modes of speciation in nature, based on the extent to which speciating populations are isolated from one another: allopatric, peripatric, parapatric, and sympatric. Speciation may also be induced artificially, through animal husbandry, agriculture, or laboratory experiments. Whether genetic drift is a minor or major contributor to speciation is the subject matter of much ongoing discussion.

Modes of Speciation

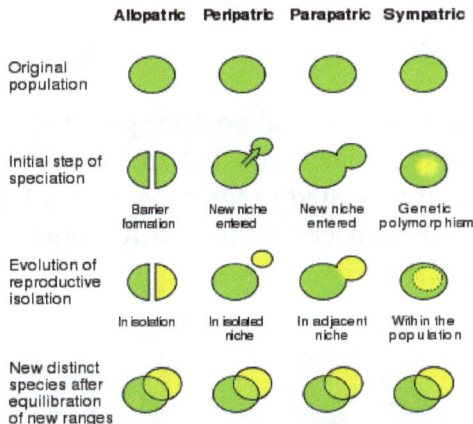

Comparison of allopatric, peripatric, parapatric and sympatric speciation

All forms of natural speciation have taken place over the course of evolution; however, debate persists as to the relative importance of each mechanism in driving biodiversity.

One example of natural speciation is the diversity of the three-spined stickleback, a marine fish that, after the last glacial period, has undergone speciation into new fresh-water colonies in isolated lakes and streams. Over an estimated 10,000 generations, the sticklebacks show structural differences that are greater than those seen between different genera of fish including variations in fins, changes in the number or size of their bony plates, variable jaw structure, and color differences.

Allopatric

During allopatric (from the ancient Greek *allos*, "other" + Greek *patrā*, "fatherland") spe-ciation, a population splits into two geographically isolated populations (for example, by habitat fragmentation due to geographical change such as mountain formation). The isolated populations then undergo genotypic or phenotypic divergence as: (a) they be-come subjected to dissimilar selective pressures; (b) they independently undergo genetic drift; (c) different mutations arise in the two populations. When the populations come back into contact, they have evolved such that they are reproductively isolated and are no longer capable of exchanging genes. Island genetics is the term associated with the tendency of small, isolated genetic pools to produce unusual traits. Examples include insular dwarfism and the radical changes among certain famous island chains, for ex-ample on Komodo. The Galápagos Islands are particularly famous for their influence on Charles Darwin. During his five weeks there he heard that Galápagos tortoises could be identified by island, and noticed that finches differed from one island to another, but it was only nine months later that he reflected that such facts could show that species were changeable. When he returned to England, his speculation on evolution deepened after experts informed him that these were separate species, not just varieties, and famously that other differing Galápagos birds were all species of finches. Though the finches were less important for Darwin, more recent research has shown the birds now known as Dar-win's finches to be a classic case of adaptive evolutionary radiation.

Peripatric

In peripatric speciation, a subform of allopatric speciation, new species are formed in isolated, smaller peripheral populations that are prevented from exchanging genes with the main population. It is related to the concept of a founder effect, since small populations often undergo bottlenecks. Genetic drift is often proposed to play a signif-icant role in peripatric speciation.

Case Studies:

- Mayr bird fauna

- The Australian bird *Petroica multicolor*

- Reproductive isolation occurs in populations of *Drosophila* subject to population bottlenecking

Parapatric

In parapatric speciation, there is only partial separation of the zones of two diverging populations afforded by geography; individuals of each species may come in contact or cross habitats from time to time, but reduced fitness of the heterozygote leads to selection for behaviours or mechanisms that prevent their interbreeding. Parapatric speciation is modelled on continuous variation within a "single," connected habitat acting as a source of natural selection rather than the effects of isolation of habitats produced in peripatric and allopatric speciation.

Parapatric speciation may be associated with differential landscape-dependent selection. Even if there is a gene flow between two populations, strong differential selection may impede assimilation and different species may eventually develop. Habitat differences may be more important in the development of reproductive isolation than the isolation time. Caucasian rock lizards *Darevskia rudis*, *D. valentini* and *D. portschinskii* all hybridize with each other in their hybrid zone; however, hybridization is stronger between *D. portschinskii* and *D. rudis*, which separated earlier but live in similar habitats than between *D. valentini* and two other species, which separated later but live in climatically different habitats.

Ecologists refer to parapatric and peripatric speciation in terms of ecological niches. A niche must be available in order for a new species to be successful. Ring species such as *Larus* gulls have been claimed to illustrate speciation in progress, though the situation may be more complex. The grass *Anthoxanthum odoratum* may be starting parapatric speciation in areas of mine contamination.

Sympatric

Freshwater angelfish, a cichlid

Sympatric speciation refers to the formation of two or more descendant species from a single ancestral species all occupying the same geographic location.

Often-cited examples of sympatric speciation are found in insects that become dependent on different host plants in the same area. However, the existence of sympatric speciation as a mechanism of speciation remains highly debated.

The best illustrated example of sympatric speciation is that of the cichlids of East Africa inhabiting the Rift Valley lakes, particularly Lake Victoria, Lake Malawi and Lake Tanganyika. There are over 800 described species, and according to estimates, there could be well over 1,600 species in the region. Their evolution is cited as an example of both natural and sexual selection. A 2008 study suggests that sympatric speciation has occurred in Tennessee cave salamanders. Sympatric speciation driven by ecological factors may also account for the extraordinary diversity of crustaceans living in the depths of Siberia's Lake Baikal.

Budding speciation has been proposed as a particular form of sympatric speciation, whereby small groups of individuals become progressively more isolated from the ancestral stock by breeding preferentially with one another. This type of speciation would be driven by the conjunction of various advantages of inbreeding such as the expression of advantageous recessive phenotypes, reducing the recombination load, and reducing the cost of sex

Rhagoletis pomonella

The hawthorn fly (*Rhagoletis pomonella*), also known as the apple maggot fly, appears to be undergoing sympatric speciation. Different populations of hawthorn fly feed on different fruits. A distinct population emerged in North America in the 19th century some time after apples, a non-native species, were introduced. This apple-feeding population normally feeds only on apples and not on the historically preferred fruit of hawthorns. The current hawthorn feeding population does not normally feed on apples. Some evidence, such as that six out of thirteen allozyme loci are different, that hawthorn flies mature later in the season and take longer to mature than apple flies; and that there is little evidence of interbreeding (researchers have documented a 4-6% hybridization rate) suggests that sympatric speciation is occurring.

Reinforcement

Reinforcement, also called the *Wallace effect*, is the process by which natural selection increases reproductive isolation. It may occur after two populations of the same species are separated and then come back into contact. If their reproductive isolation was complete, then they will have already developed into two separate incompatible species. If their reproductive isolation is incomplete, then further mating between the populations will produce hybrids, which may or may not be fertile. If the hybrids are infertile, or fertile but less fit than their ancestors, then there will be further reproductive isolation and speciation has essentially occurred (e.g., as in horses and donkeys.)

The reasoning behind this is that if the parents of the hybrid offspring each have naturally selected traits for their own certain environments, the hybrid offspring will bear traits from both, therefore would not fit either ecological niche as well as either parent. The low fitness of the hybrids would cause selection to favor assortative mating, which would control hybridization. This is sometimes called the Wallace effect after the evolutionary biologist Alfred Russel Wallace who suggested in the late 19th century that it might be an important factor in speciation. Conversely, if the hybrid offspring are more fit than their ancestors, then the populations will merge back into the same species within the area they are in contact.

Reinforcement favoring reproductive isolation is required for both parapatric and sympatric speciation. Without reinforcement, the geographic area of contact between different forms of the same species, called their "hybrid zone," will not develop into a boundary between the different species. Hybrid zones are regions where diverged populations meet and interbreed. Hybrid offspring are very common in these regions, which are usually created by diverged species coming into secondary contact. Without reinforcement, the two species would have uncontrollable inbreeding. Reinforcement may be induced in artificial selection experiments as described below.

Ecological and Parallel Speciation

Ecological selection is "the interaction of individuals with their environment during resource acquisition". Natural selection is inherently involved in the process of speciation, whereby, "under ecological speciation, populations in different environments, or populations exploiting different resources, experience contrasting natural selection pressures on the traits that directly or indirectly bring about the evolution of reproductive isolation". Evidence for the role ecology plays in the process of speciation exists. Studies of stickleback populations support ecologically-linked speciation arising as a by-product, alongside numerous studies of parallel speciation.

Parallel speciation is where "greater reproductive isolation repeatedly evolves between independent populations adapting to contrasting environments than between independent populations adapting to similar environments". It is established that ecological speciation occurs and with much of the evidence, "...accumulated from top-down studies of adaptation and reproductive isolation".

Sexual Selection

It is widely appreciated that sexual selection could drive speciation in many clades, independently of natural selection. However the term "speciation", in this context, tends to be used in two different, but not mutually exclusive senses. The first and most commonly used sense refers to the "birth" of new species. That is, the splitting of an existing species into two separate species, or the budding off of a new species from a parent species, both driven by a biological "fashion fad" (a preference for a feature, or features, in one or both sexes, that do not necessarily have any adaptive qualities). In the second sense, "speciation" refers the wide-spread tendency of sexual creatures to be grouped into clearly defined species, rather than forming a continuum of phenotypes both in time and space - which would be the more obvious or logical consequence of natural selection. This was indeed recognized by Darwin as problematic, and included in his *On the Origin of Species* (1859), under the heading "Difficulties with the Theory". There are several suggestions as to how mate choice might play a significant role in resolving Darwin's dilemma.

Artificial Speciation

European mouflon (*Ovis aries musimon*)

New species have been created by domesticated animal husbandry, but the initial dates and methods of the initiation of such species are not clear. Often, the domestic counterpart of the wild ancestor can still interbreed and produce fertile offspring as in the case of domestic cattle, that can be considered the same species as several varieties of wild ox, gaur, yak, etc., or domestic sheep that can interbreed with the mouflon.

The best-documented creations of new species in the laboratory were performed in the late 1980s. William R. Rice and George W. Salt bred *Drosophila melanogaster* fruit flies using a maze with three different choices of habitat such as light/dark and wet/

dry. Each generation was placed into the maze, and the groups of flies that came out of two of the eight exits were set apart to breed with each other in their respective groups. After thirty-five generations, the two groups and their offspring were isolated reproductively because of their strong habitat preferences: they mated only within the areas they preferred, and so did not mate with flies that preferred the other areas. The history of such attempts is described by Rice and Elen E. Hostert (1993).

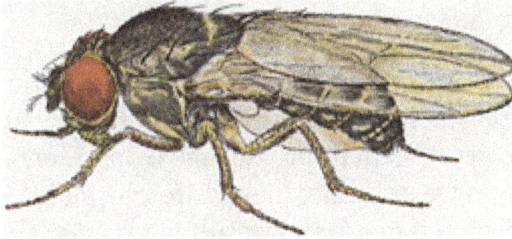

Male *Drosophila pseudoobscura*

Diane Dodd used a laboratory experiment to show how reproductive isolation can evolve in *Drosophila pseudoobscura* fruit flies after several generations by placing them in different media, starch- and maltose-based media.

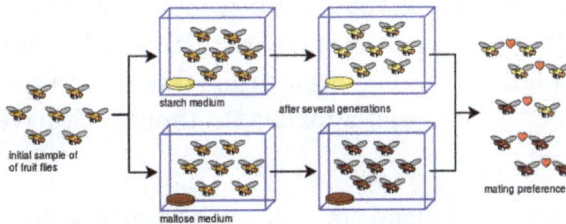

Dodd's experiment has been easy for many others to replicate, including with other kinds of fruit flies and foods. Research in 2005 has shown that this rapid evolution of reproductive isolation may in fact be a relic of infection by *Wolbachia* bacteria.

Alternatively, these observations are consistent with the notion that sexual creatures are inherently reluctant to mate with individuals whose appearance or behavior is different from the norm. The risk that such deviations are due to heritable maladaptations is very high. Thus, if a sexual creature, unable to predict natural selection's future direction, is conditioned to produce the fittest offspring possible, it will avoid mates with unusual habits or features. Sexual creatures will then inevitably tend to group themselves into reproductively isolated species.

Genetics

Few speciation genes have been found. They usually involve the reinforcement process of late stages of speciation. In 2008, a speciation gene causing reproductive isolation was reported. It causes hybrid sterility between related subspecies. The order of speciation of three groups from a common ancestor may be unclear or unknown; a collection of three such species is referred to as a "trichotomy."

Speciation via Polyploidization

Polyploidy is a mechanism that has caused many rapid speciation events in sympatry because offspring of, for example, tetraploid x diploid matings often result in triploid sterile progeny. However, not all polyploids are reproductively isolated from their parental plants, and gene flow may still occur for example through triploid hybrid x diploid matings that produce tetraploids, or matings between meiotically unreduced gametes from diploids and gametes from tetraploids.

It has been suggested that many of the existing plant and most animal species have undergone an event of polyploidization in their evolutionary history. Reproduction of successful polyploid species is sometimes asexual, by parthenogenesis or apomixis, as for unknown reasons many asexual organisms are polyploid. Rare instances of polyploid mammals are known, but most often result in prenatal death.

Hybrid Speciation

Hybridization between two different species sometimes leads to a distinct phenotype. This phenotype can also be fitter than the parental lineage and as such natural selection may then favor these individuals. Eventually, if reproductive isolation is achieved, it may lead to a separate species. However, reproductive isolation between hybrids and their parents is particularly difficult to achieve and thus hybrid speciation is considered an extremely rare event. The mariana mallard is thought to have arisen from hybrid speciation.

Hybridization is an important means of speciation in plants, since polyploidy (having more than two copies of each chromosome) is tolerated in plants more readily than in animals. Polyploidy is important in hybrids as it allows reproduction, with the two different sets of chromosomes each being able to pair with an identical partner during meiosis. Polyploids also have more genetic diversity, which allows them to avoid inbreeding depression in small populations.

Hybridization without change in chromosome number is called homoploid hybrid speciation. It is considered very rare but has been shown in *Heliconius* butterflies and sunflowers. Polyploid speciation, which involves changes in chromosome number, is a more common phenomenon, especially in plant species.

Gene Transposition

Theodosius Dobzhansky, who studied fruit flies in the early days of genetic research in 1930s, speculated that parts of chromosomes that switch from one location to another might cause a species to split into two different species. He mapped out how it might be possible for sections of chromosomes to relocate themselves in a genome. Those mobile sections can cause sterility in inter-species hybrids, which can act as a speciation pressure. In theory, his idea was sound, but scientists long debated whether it actually happened in nature. Eventually a competing theory involving the gradual accumulation of mutations was shown to occur in nature so often that geneticists largely dismissed the moving gene hypothesis.

However, 2006 research shows that jumping of a gene from one chromosome to another can contribute to the birth of new species. This validates the reproductive isolation mechanism, a key component of speciation.

Historical Background

In addressing the question of the origin of species, there are two key issues: (1) what are the evolutionary mechanisms of speciation, and (2) what accounts for the separateness and individuality of species in the biota? Since Charles Darwin's time, efforts to understand the nature of species have primarily focused on the first aspect, and it is now widely agreed that the critical factor behind the origin of new species is reproductive isolation. Next we focus on the second aspect of the origin of species.

Darwin's Dilemma: Why do species exist?

In *On the Origin of Species* (1859), Darwin interpreted biological evolution in terms of natural selection, but was perplexed by the clustering of organisms into species. Chapter 6 of Darwin's book is entitled "Difficulties of the Theory." In discussing these "difficulties" he noted "Firstly, why, if species have descended from other species by insensibly fine gradations, do we not everywhere see innumerable transitional forms? Why is not all nature in confusion instead of the species being, as we see them, well defined?" This dilemma can be referred to as the absence or rarity of transitional varieties in habitat space.

Another dilemma, related to the first one, is the absence or rarity of transitional varieties in time. Darwin pointed out that by the theory of natural selection "innumerable transitional forms must have existed," and wondered "why do we not find them embedded in countless numbers in the crust of the earth." That clearly defined species actually do exist in nature in both space and time implies that some fundamental feature of natural selection operates to generate and maintain species.

The Effect of Sexual Reproduction on Species Formation

It has been argued that the resolution of Darwin's first dilemma lies in the fact that out-crossing sexual reproduction has an intrinsic cost of rarity. The cost of rarity arises as follows. If, on a resource gradient, a large number of separate species evolve, each exquisitely adapted to a very narrow band on that gradient, each species will, of necessity, consist of very few members. Finding a mate under these circumstances may present difficulties when many of the individuals in the neighborhood belong to other species. Under these circumstances, if any species' population size happens, by chance, to increase (at the expense of one or other of its neighboring species, if the environment is saturated), this will immediately make it easier for its members to find sexual partners. The members of the neighboring species, whose population sizes have decreased, experience greater difficulty in finding mates, and therefore form pairs less frequently than the larger species. This has a snowball effect, with large species growing at the expense of the smaller, rarer species, eventually driving them to extinction. Eventually, only a few species remain, each distinctly different from the other. The cost of rarity not only involves the costs of failure to find a mate, but also indirect costs such as the cost of communication in seeking out a partner at low population densities.

African pygmy kingfisher, showing coloration shared by all adults of that species to a high degree of fidelity.

Rarity brings with it other costs. Rare and unusual features are very seldom advantageous. In most instances, they indicate a (non-silent) mutation, which is almost certain to be deleterious. It therefore behooves sexual creatures to avoid mates sporting rare or unusual features. Sexual populations therefore rapidly shed rare or peripheral phenotypic features, thus canalizing the entire external appearance, as illustrated in the accompanying illustration of the African pygmy kingfisher, *Ispidina picta*. This remarkable uniformity of all the adult members of a sexual species has stimulated the proliferation of field guides on birds, mammals, reptiles, insects, and many other taxa, in which a species can be described with a single illustration (or two, in the case of sexual dimorphism). Once a population has become as homogeneous in appearance as is typical of most species (and is illustrated in the photograph of the African pygmy kingfisher), its members will avoid mating with members of other populations that look different from themselves. Thus, the avoidance of mates displaying rare and unusual phenotypic features inevitably leads to reproductive isolation, one of the hallmarks of speciation.

In the contrasting case of organisms that reproduce asexually, there is no cost of rarity; consequently, there are only benefits to fine-scale adaptation. Thus, asexual organisms very frequently show the continuous variation in form (often in many different directions) that Darwin expected evolution to produce, making their classification into "species" (more correctly, morphospecies) very difficult.

Rates of Speciation

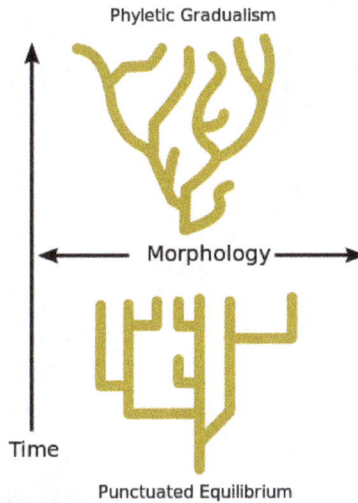

Phyletic Gradualism

Morphology

Time

Punctuated Equilibrium

Phyletic gradualism, above, consists of relatively slow change over geological time. Punctuated equilibrium, bottom, consists of morphological stability and rare, relatively rapid bursts of evolutionary change.

There is debate as to the rate at which speciation events occur over geologic time. While some evolutionary biologists claim that speciation events have remained relatively constant and gradual over time (known as "Phyletic gradualism"), some palaeontologists such as Niles Eldredge and Stephen Jay Gould have argued that species usually remain unchanged over long stretches of time, and that speciation occurs only over relatively brief intervals, a view known as *punctuated equilibrium.*

Punctuated Evolution

Evolution can be extremely rapid, as shown in the creation of domesticated animals and plants in a very short geological space of time, spanning only a few tens of thousands of years. Maize (*Zea mays*), for instance, was created in Mexico in only a few thousand years, starting about 7,000 to 12,000 years ago. This raises the question of why the long term rate of evolution is far slower than is theoretically possible.

Plants and domestic animals can differ markedly from their wild ancestors

Top: wild teosinte; middle: maize-teosinte hybrid; bottom: maize

Ancestral wild cabbage

Domesticated cauliflower

Ancestral Prussian carp

Domestic goldfish

Ancestral mouflon

Evolution is imposed on species or groups. It is not planned or striven for in some Lamarckist way. The mutations on which the process depends are random events, and, except for the "silent mutations" which do not affect the functionality or appearance of the carrier, are thus usually disadvantageous, and their chance of proving to be useful in the future is vanishingly small. Therefore, while a species or group might benefit from being able to adapt to a new environment by accumulating a wide range of genetic variation, this is to the detriment of the *individuals* who have to carry these mutations until a small, unpredictable minority of them ultimately contributes to such an adaptation. Thus, the *capability* to evolve would require group selection, a concept discredited by (for example) George C. Williams, John Maynard Smith and Richard Dawkins as selectively disadvantageous to the individual.

The resolution to Darwin's second dilemma might thus come about as follows:

If sexual individuals are disadvantaged by passing mutations on to their offspring, they will avoid mutant mates with strange or unusual characteristics. Mutations that affect the external appearance of their carriers will then rarely be passed on to the next and subsequent generations. They would therefore seldom be tested by natural selection. Evolution is, therefore, effectively halted or slowed down considerably. The only mutations that can accumulate in a population, on this punctuated equilibrium view, are ones that have no noticeable effect on the outward appearance and functionality of their bearers (i.e., they are "silent" or "neutral mutations," which can be, and are, used to trace the relatedness and age of populations and species.) This argument implies that evolution can only occur if mutant mates cannot be avoided, as a result of a severe scarcity of potential mates. This is most likely to occur in small, isolated communities. These occur most commonly on small islands, in remote valleys, lakes, river systems, or caves, or during the aftermath of a mass extinction. Under these circumstances, not only is the choice of mates severely restricted but population bottlenecks, founder effects, genetic drift and inbreeding cause rapid, random changes in the isolated population's genetic composition. Furthermore, hybridization with a related species trapped in the same isolate might introduce additional genetic changes. If an isolated population such as this survives its genetic upheavals, and subsequently expands into an unoccupied niche, or into a niche in which it has an advantage over its competitors, a new species, or subspecies, will have come in being. In geological terms this will be an abrupt event. A resumption of avoiding mutant mates will thereafter result, once again, in evolutionary stagnation.

In apparent confirmation of this punctuated equilibrium view of evolution, the fossil record of an evolutionary progression typically consists of species that suddenly appear, and ultimately disappear, hundreds of thousands or millions of years later, without any change in external appearance. Graphically, these fossil species are represented by horizontal lines, whose lengths depict how long each of them existed. The horizontality of the lines illustrates the unchanging appearance of each of the fossil species depicted on the graph. During each species' existence new species appear at random intervals, each also lasting many hundreds of thousands of years before disappearing without a change in appearance. The exact relatedness of these concurrent species is generally impossible to determine. This is illustrated in the diagram depicting the distribution of hominin species through time since the hominins separated from the line that led to the evolution of our closest living primate relatives, the chimpanzees.

For similar evolutionary time lines see, for instance, the paleontological list of African dinosaurs, Asian dinosaurs, the Lampriformes and Amiiformes.

Peripatric Speciation

Peripatric speciation is a mode of speciation in which a new species is formed from an isolated peripheral population. Since peripatric speciation resembles allopatric speciation, in that populations are isolated and prevented from exchanging genes, it can often be difficult to distinguish between them. Nevertheless, the primary characteristic of peripatric speciation proposes that one of the populations is much smaller than the other.

The terms peripatric and peripatry are often used in biogeography, referring to organisms whose ranges are closely adjacent but do not overlap, being separated where these organisms do not occur—for example on an oceanic island compared to the mainland. Such organisms are usually closely related (e.g. sister species); their distribution being the result of peripatric speciation. An alternative model of peripatric speciation, centrifugal speciation, posits that a species' population experiences periods of geographic range expansion followed by shrinking periods, leaving behind small isolated populations on the periphery of the main population. Other models have involved the effects of sexual selection on limited population sizes.

The existence of peripatric speciation is supported by laboratory experiments and empirical evidence. Scientists observing the patterns of a species biogeographic distribution and its phylogenetic relationships are able to reconstruct the historical process by which they diverged. Further, oceanic islands are often the subject of peripatric speciation research due to their isolated habitats—with the Hawaiian Islands widely represented in much of the scientific literature.

History

Peripatric speciation was first proposed by Ernst Mayr in 1954 and fully theoretically modeled in his 1982. It is related to the founder effect, where small living populations may undergo selection bottlenecks. The founder effect is based on models that suggest peripatric speciation can occur by the interaction of selection and genetic drift, which may play a significant role. In 1976 and 1980, the Kaneshiro model of peripatric speciation was developed by Kenneth Y. Kaneshiro which focused on sexual selection as a driver for speciation during population bottlenecks.

Models

Peripatric speciation models are identical to models of vicariance (allopatric speciation). Requiring both geographic separation and time, speciation can result as a predictable byproduct. Peripatry can be distinguished from allopatric speciation by three features: 1) the size of the isolated population, 2) strong selection caused by the dispersal and colonization of novel environments, and 3) the effects of genetic drift on small populations. The peripatric model results in, what have been called, progenitor-derivative species pairs, whereby the derivative species (the peripherally isolated population)—geographically and genetically isolated from the progenitor species—diverges.

The size of a population is important because individuals colonizing a new habitat likely contain only a small sample of the genetic variation of the original population. This promotes divergence due to strong selective pressures, leading to the rapid fixation of an allele within the descendant population. This gives rise to the potential for genetic incompatibilities to evolve. These incompatibilities cause reproductive isolation, giving rise to—sometimes rapid—speciation events.

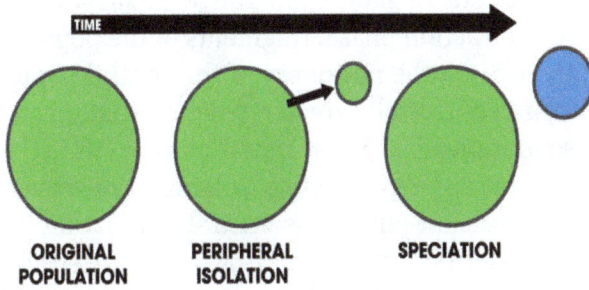

a: Peripatric speciation

Speciation under the peripatric model invokes two important predictions, namely that geological or climactic changes cause populations to become locally fragmented (or regionally when considering allopatric speciation), and that isolated population's reproductive traits evolve enough as to prevent interbreeding upon potential secondary contact. The phylogenetic signature of this model is that the central population remains pleisomorphic, while the peripheral isolates become apomorphic.

One possible consequence of peripatric speciation is that a geographically widespread ancestral species becomes paraphyletic, thereby becoming a paraspecies. The concept of a paraspecies is therefore a logical consequence of the evolutionary species concept, by which one species give rise to a daughter species.

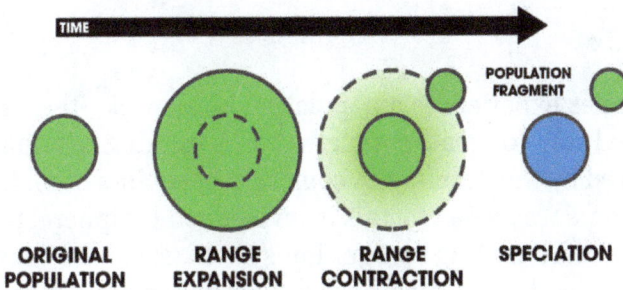

b: Centrifugal speciation

Diagrams representing the process of peripatric and centrifugal speciation. In peripatry, a small population becomes isolated on the periphery of the central population evolving reproductive isolation (blue) due to reduced gene flow. In centrifugal speciation, an original population (green) range expands and contracts, leaving an isolated fragment population behind. The central population (changed to blue) evolves reproductive isolation in contrast to peripatry.

Centrifugal Speciation

William Louis Brown, Jr. proposed an alternative model of peripatric speciation in 1957 called centrifugal speciation. This model contrasts with peripatric speciation by virtue of the origin of the genetic novelty that leads to reproductive isolation. A population of a species experiences periods of geographic range expansion followed by periods of contraction. During the contraction phase, fragments of the population become isolated as small refugial populations on the periphery of the central population (see figure b). Because of the large size and potentially greater genetic variation within the central population, mutations arise more readily. These mutations are left in the isolated peripheral populations, whereby, promoting reproductive isolation. Consequently, Brown suggested that during another expansion phase, the central population would overwhelm the peripheral populations, hindering speciation. However, if the species finds a specialized ecological niche, the two may coexist. The phylogenetic signature of this model is that the central population becomes derived, while the peripheral isolates stay pleisomorphic.

Centrifugal speciation has been largely ignored in the scientific literature, often dominated by the traditional model of peripatric speciation. Despite this, Brown cited a wealth of evidence to support his model, of which has not yet been refuted.

Peromyscus polionotus and *P. melanotis* (the peripherally isolated species from the central population of *P. maniculatus*) arose via the centrifugal speciation model. Centrifugal speciation may have taken place in tree kangaroos, South American frogs (*Ceratophrys*), shrews (*Crocidura*), and primates (*Presbytis melalophos*). John C. Briggs associates centrifugal speciation with centers of origin, contending that the centrifugal model is better supported by the data, citing species patterns from the proposed 'center of origin' within the Indo-West Pacific

Kaneshiro Model

When a sexual species experiences a population bottleneck—that is, when the genetic variation is reduced due to small population size—mating discrimination among females may be altered by the decrease in courtship behaviors of males. Sexual selection pressures may become weakened by this in an isolated peripheral population, and as a by-product of the altered mating recognition system, secondary sexual traits may appear. Eventually, a growth in population size paired with novel female mate preferences will give rise to reproductive isolation from the main population-thereby completing the peripatric speciation process.

Support for this model comes from experiments and observation of species that exhibit asymmetric mating patterns such as the Hawaiian *Drosophila* species or the Hawaiian cricket *Laupala*. However, this model has not been entirely supported by experiments, and therefore, it may not represent a plausible process of peripatric speciation that takes place in nature.

Evidence

Laboratory Experiments

Peripatric speciation has been researched in both laboratory studies and nature. Coyne and Orr in *Speciation* suggest that most laboratory studies of allopatric speciation are also examples of peripatric speciation due to their small population sizes and the inevitable divergent selection that they undergo.

Much of the laboratory research concerning peripatry is inextricably linked to founder effect research. Coyne and Orr conclude that selection's role in speciation is well established, whereas genetic drift's role is unsupported by experimental and field data—suggesting that founder-effect speciation does not occur. Nevertheless, a great deal of research has been conducted on the matter, and one study conducted involving bottleneck populations of *Drosophila pseudoobscura* found evidence of isolation after a single bottleneck.

Below is a non-exhaustive table of laboratory experiments focused explicitly on peripatric speciation. Most of the studies also conducted experiments on vicariant speciation as well. The "replicates" column signifies the number of lines used in the experiment—that is, how many independent populations were used (not the population size or the number of generations performed).

Laboratory experiments of peripatric speciation		
Species	**Replicates**	**Year**
Drosophila adiastola	1	1979
Drosophila silvestris	1	1980
Drosophila pseudoobscura	8	1985
Drosophila simulans	8	1985
Musca domestica	6	1991
Drosophila pseudoobscura	42	1993
Drosophila melanogaster	50	1998
Drosophila melanogaster	19; 19	1999
Drosophila grimshawi	1	N/A

Species Patterns on Islands and Archipelagos

Island species provide direct evidence of speciation occurring peripatrically in such that, "the presence of endemic species on oceanic islands whose closest relatives inhabit a nearby continent" must have originated by a colonization event. Comparative phylogeography of oceanic archipelagos shows consistent patterns of sequential colonization and speciation along island chains, most notably on the Azores islands, Canary Islands, Society Islands, Marquesas Islands, Galápagos Islands, Austral Islands, and the Hawaiian Islands—all of which express geological patterns of spatial isolation and, in some cases, linear arrangement.

Hawaiian Archipelago

Drosophila species on the Hawaiian archipelago have helped researchers understand speciation processes in great detail. It is well established that *Drosophila* has undergone an adaptive radiation into hundreds of endemic species on the Hawaiian island chain; originating from a single common ancestor (supported from molecular analysis). Studies consistently find that colonization of each island occurred from older to younger islands, and in *Drosophila*, speciating peripatrically at least fifty percent of the time. In conjunction with *Drosophila*, Hawaiian lobeliads (*Cyanea*) have also undergone an adaptive radiation, with upwards of twenty-seven percent of extant species arising after new island colonization—exemplifying peripatric speciation—once again, occurring in the old-to-young island direction.

Other endemic species in Hawaii also provide evidence of peripatric speciation such as the endemic flightless crickets (*Laupala*). It has been estimated that, "17 species out of 36 well-studied cases of [*Laupala*] speciation were peripatric". Plant species in genera's such as *Dubautia*, *Wilkesia*, and *Argyroxiphium* have also radiated along the archipelago.

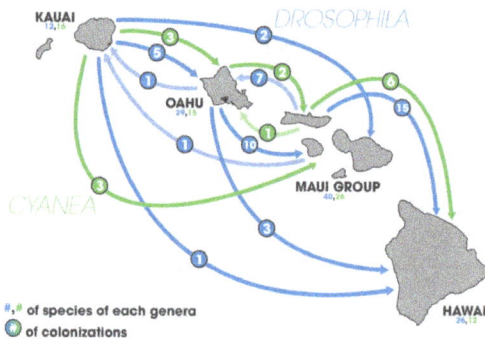

Figure c: Colonization events of species from the genus *Cyanea* (green) and species from the genus *Drosophila* (blue) on the Hawaiian island chain. Islands age from left to right, (Kauai being the oldest and Hawaii being the youngest). Speciation arises peripatrically as they spatiotemporally colonize new islands along the chain. Lighter blue and green indicate colonization in the reverse direction from young-to-old.

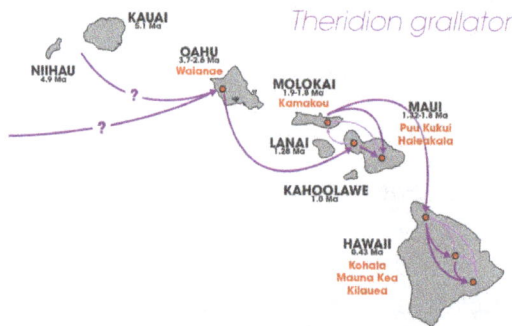

Figure d: A map of the Hawaiian archipelago showing the colonization routes of *Theridion grallator* superimposed. Purple lines indicate colonization occurring in conjunction with island age where light purple indicates backwards colonization. *T. grallator* is not present on Kauai or Niihau so colonization may have occurred from there, or the nearest continent.

Tetragnatha spiders have also speciated peripatrically on the Hawaiian islands, Numerous arthropods have been documented existing in patterns consistent with the geologic evolution of the island chain, in such that, phylogenetic reconstructions find younger species inhabiting the geologically younger islands and older species inhabiting the older islands (or in some cases, ancestors date back to when islands currently below sea level were exposed). Spiders such such as those from the genus *Orsonwelles* exhibit patterns compatible with the old-to-young geology. Other endemic genera such as *Argyrodes* have been shown to have speciated along the island chain. *Pagiopalus*, *Pedinopistha*, and part of the Thomisidae family have adaptively radiated along the island chain, as well as the Lycosidae family of wolf spiders.

A host of other Hawaiian endemic arthropod species and genera have had their speciation and phylogeographical patterns studied: the *Drosophila grimshawi* species complex, damselflies (*Megalagrion xanthomelas* and *Megalagrion pacificum*), *Doryonychus raptor*, *Littorophiloscia hawaiiensis*, *Anax strenuus*, *Nesogonia blackburni*, *Theridion grallator* (see figure c), *Vanessa tameamea*, *Hyalopeplus pellucidus*, *Coleotichus blackburniae*, *Labula*, *Hawaiioscia*, *Banza*, *Caconemobius*, *Eupethicea*, *Ptycta*, *Megalagrion*, *Prognathogryllus*, *Nesosydne*, *Cephalops*, *Trupanea*, and the tribe Platynini—all suggesting repeated radiations among the islands.

Other animals besides insects show this same pattern such as the Hawaiian amber snail (*Succinea caduca*), and 'Elepaio flycatchers.

Other Islands

Phylogenetic studies of a species of crab spider (*Misumenops rapaensis*) in the genus Thomisidae located on the Austral Islands have established the, "sequential colonization of [the] lineage down the Austral archipelago toward younger islands". Interestingly, *M. rapaensis* has been traditionally thought of as a single species; whereas this particular study found distinct genetic differences corresponding to the sequential age of the islands.

Species Patterns on Continents

Figure e: Satellite image of the Noel Kempff Mercado National Park in Bolivia (outlined in red). The red arrow indicates the location of the isolated forest fragment.

The occurrence of peripatry on continents is more difficult to detect due to the possibility of vicariant explanations being equally likely. However, studies concerning the Californian plant species *Clarkia biloba* and *C. biloba* strongly suggest a peripatric origin. In addition, a great deal of research has been conducted on several species of land snails involving chirality that suggests peripatry (with some authors noting other possible interpretations).

A study by Lucinda P. Lawson et al. found evidence for the occurrence of peripatric speciation in the montane spiny throated reed frog species complex (genus: *Hyperolius*). Lawson maintains that the species' geographic ranges within the Eastern Afromontane Biodiversity Hotspot support a peripatric model that is driving speciation; suggesting that this mode of speciation may play a significant role in "highly fragmented ecosystems".

The chestnut-tailed antbird is located within the Serrania de Huanchaca in Bolivia. Within this region exists an fringe patch of forest (see figure e) estimated to have been isolated for 1000–3000 years. Researchers Nathalie Seddon and Joseph A. Tobias found significant song divergence in the population of birds that reside in this isolated patch. They concluded that this measured divergence is evidence of an "early step" in the process of peripatric speciation and "may partly explain the dramatic diversification of suboscines in Amazonia".

In a study of the phylogeny and biogeography of the land snail genus *Monacha*, the species *M. ciscaucasica* is thought to have speciated peripatrically from a population of *M. roseni*. In addition, *M. claussi* consists of a small population located on the peripheral of the much larger range of *M. subcarthusiana* suggesting that it also arose by peripatric speciation.

Red spruce (*Picea rubens*) has arisen from an isolated population of black spruce (*Picea mariana*). During the Pleistocene, a population of black spruce became geographically isolated, likely due to glaciation. The geographic range of the black spruce is much larger than the red spruce. The red spruce has significantly lower genetic diversity in both its DNA and its mitochondrial DNA than the black spruce. Furthermore, the genetic variation of the red spruce has no unique mitochondrial haplotypes, only subsets of those in the black spruce; suggesting that the red spruce speciated peripatrically from the black spruce population. It is thought that the entire *Picea* genus in North America has been diversified by the process of peripatric speciation, as numerous pairs of closely related species in the genus have smaller southern population ranges; and those with overlapping ranges often exhibit weak reproductive isolation.

Using a phylogeographic approach paired with ecological niche models (i.e. prediction and identification of expansion and contraction species ranges into suitable habitats based on current ecological niches, correlated with fossil and molecular data), researchers found that the prairie dog species *Cynomys mexicanus* speciated peripatrically from *Cynomys ludovicianus* approximately 230,000 years ago. North American glacial cycles promoted range expansion and contraction of the prairie dogs, leading to the isolation of a relic population in a refugium located in the present day Coahuila, Mexico. This distribution and paleobiogeographic pattern correlates with other species expressing similar biographic range patterns such as with the *Sorex cinereus* complex.

Parapatric Speciation

In parapatric speciation, two subpopulations of a species evolve reproductive isolation from one another while continuing to exchange genes. This mode of speciation has three distinguishing characteristics: 1) mating occurs non-randomly, 2) gene flow occurs unequally, and 3) populations exist in either continuous or discontinuous geographic ranges. This distribution pattern may be the result of unequal dispersal, incomplete geographical barriers, or divergent expressions of behavior, among other things. Parapatric speciation predicts that hybrid zones will often exist at the junction between the two populations.

In biogeography, the terms parapatric and parapatry are often used to describe the relationship between organisms whose ranges do not significantly overlap but are immediately adjacent to each other; they do not occur together except in a narrow contact zone. Parapatry is a geographical distribution opposed to sympatry (same area) and allopatry or peripatry (two similar cases of distinct areas).

Various "forms" of parapatry have been proposed and are discussed below. Coyne and Orr in *Speciation* categorise these forms into three groups: clinal (environmental gradients), "stepping-stone" (discrete populations), and stasipatric speciation in concordance with most of the parapatric speciation literature. Henceforth, the models are subdivided following a similar format.

Charles Darwin was the first to propose this mode of speciation. It was not until 1930 when Ronald Fisher published *The Genetical Theory of Natural Selection* where he outlined a verbal theoretical model of clinal speciation. In 1981, Joseph Felsenstein proposed an alternative, "discrete population" model (the "stepping-stone model). Since Darwin, a great deal of research has been conducted on parapatric speciation—concluding that its mechanisms are theoretically plausible "and has most certainly occurred in nature".

Models

Mathematical models, laboratory studies, and observational evidence supports the existence of parapatric speciation's occurrence in nature. The qualities of parapatry imply a partial extrinsic barrier during divergence; thus leading to a difficulty in determining whether this mode of speciation actually occurred, or if an alternative mode (notably, allopatric speciation) can explain the data. This problem poses the unanswered question as to its overall frequency in nature.

Parapatric speciation can be understood as a level of gene flow between populations where $m=0$ in allopatry (and peripatry), $m=0.5$ in sympatry, and midway between the two in parapatry. Intrinsic to this, parapatry covers the entire continuum; represented as $0 < m < 0.5$. Some biologists reject this delineation, advocating the disuse of the term "parapatric" outright, "because many different spatial distributions can result in intermediate levels of gene flow". Others champion this position and suggest the abandonment of geographic classification schemes (geographic modes of speciation) altogether.

Natural selection has been shown to be the primary driver in parapatric speciation (among other modes), and the strength of selection during divergence is often an important factor.

Parapatric speciation may also result from reproductive isolation caused by social selection: individuals interacting altruistically.

Environmental Gradients

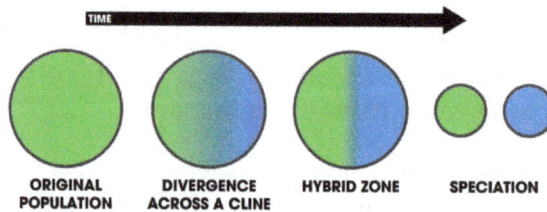

A diagram representing population subject to a selective gradient of phenotypic or genotypic frequencies (a cline). Each end of the gradient experiences different selective conditions (divergent selection). Reproductive isolation occurs upon the formation of a hybrid zone. In most cases, the hybrid zone may become eliminated due to a selective disadvantage. This effectively completes the speciation process.

Due to the continuous nature of a parapatric population distribution, population niches will often overlap, producing a continuum in the species' ecological role across an environmental gradient. Whereas in allopatric or peripatric speciation—in which geographically isolated populations may evolve reproductive isolation without gene flow—the reduced gene flow of parapatric speciation will often produce a cline in which a variation in evolutionary pressures causes a change to occur in allele frequencies within the gene pool between populations. This environmental gradient ultimately results in genetically distinct sister species.

Fisher's original conception of clinal speciation relied on—unlike most modern speciation research—the morphological species concept. With this interpretation, his verbal, theoretical model *can* effectively produce a new species; of which was subsequently confirmed mathematically. Further mathematical models have been developed to demonstrate the possibility of clinal speciation with most relying on, what Coyne and Orr assert are, "assumptions that are either restrictive or biologically unrealistic".

A mathematical model for clinal speciation was developed by Caisse and Antonovics that found evidence that, "both genetic divergence and reproductive isolation may therefore occur between populations connected by gene flow". This research supports clinal isolation comparable to a ring species (discussed below), except that the terminal geographic ends do not meet to form a ring.

Doebeli and Dieckmann developed a mathematical model that suggested that ecological contact is an important factor in parapatric speciation and that, despite gene flow acting as a barrier to divergence in the local population, disruptive selection drives assortative mating; eventually leading to a complete reduction in gene flow. This model resembles reinforcement with the exception that there is never a secondary contact event. The authors conclude that, "spatially localized interactions along environmental gradients can facilitate speciation through frequency-dependent selection and result in patterns of geographical segregation between the emerging species." However, one study by Polechová and Barton disputes these conclusions.

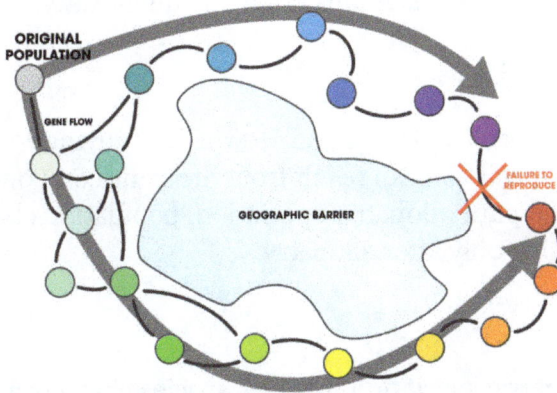

In a ring species, individuals are able to successfully reproduce (exchange genes) with members of their own species in adjacent populations occupying a suitable habitat around a geographic barrier. Individuals a the ends of the cline are unable to reproduce when they come into contact.

Ring Species

The concept of a ring species is associated with allopatric speciation as a special case; however, Coyne and Orr argue that Mayr's original conception of a ring species does not describe allopatric speciation, "but speciation occurring through the attenuation of gene flow with distance". They contend that ring species provide evidence of parapatric speciation in a non-conventional sense. They go on to conclude that:

Nevertheless, ring species are more convincing than cases of clinal isolation for showing that gene flow hampers the evolution of reproductive isolation. In clinal isolation, one can argue that reproductive isolation was caused by environmental differences that increase with distance between populations. One cannot make a similar argument for ring species because the most reproductively isolated populations occur in the *same* habitat.

Discrete Populations

Referred to as a "stepping-stone" model by Coyne and Orr, it differs by virtue of the species population distribution pattern. Populations in discrete groups undoubtedly speciate more easily than those in a cline due to more limited gene flow. This allows for a population to evolve reproductive isolation as either selection or drift overpower gene flow between the populations. The smaller the discrete population, the species will likely undergo a higher rate of parapatric speciation.

Several mathematical models have been developed to test whether this form of parapatric speciation can occur, providing theoretical possibility and supporting biological plausibility (dependent on the models parameters and their concordance with nature).

Joseph Felsenstein was the first to develop a working model. Later, Sergey Gavrilets and colleagues developed numerous analytical and dynamical models of parapatric speciation that have contributed significantly to the quantitative study of speciation.

Para-allopatric Speciation

Further concepts developed by Barton and Hewitt in studying 170 hybrid zones, suggested that parapatric speciation can result from the same components that cause allopatric speciation. Called para-allopatric speciation, populations begin diverging parapatrically, fully speciating only after allopatry.

Stasipatric Models

One variation of parapatric speciation involves species chromosomal differences. Michael J. D. White developed the stasipatric speciation model when studying Australian morabine grasshoppers (*Vandiemenella*). The chromosomal structure of sub-populations of a widespread species become underdominate; leading to fixation. Subsequently, the sub-populations expand within the species larger range, hybridizing (with sterility of the offspring) in narrow hybrid zones. Futuyama and Mayer contend that this form of parapatric speciation is untenable and that chromosomal rearrangements are unlikely to cause speciation. Nevertheless, data does support that chromosomal rearrangements can possibly lead to reproductive isolation, but it does not mean speciation results as a consequence.

Evidence

Laboratory Evidence

Very few laboratory studies have been conducted that explicitly test for parapatric speciation. However, research concerning sympatric speciation often lends support to the occurrence of parapatry. This is due to the fact that, in symaptric speciation, gene flow within a population is unrestricted; whereas in parapatric speciation, gene flow is limited—thus allowing reproductive isolation to evolve easier. Ödeen and Florin complied 63 laboaratory experiments conducted between the years 1950–2000 (many of which were discussed by Rice and Hostert previously) concerning sympatric and parapatric speciation. They contend that the laboratory evidence is more robust than often suggested, citing laboratory populations sizes as the primary shortcoming.

Observational Evidence

Parapatric speciation is very difficult to observe in nature. This is due to one primary factor: patterns of parapatry can easily be explained by an alternate mode of speciation. Particularly, documenting closely related species sharing common boundaries does not imply that parapatric speciation was the mode that created this geographic distribution pattern. Coyne and Orr assert that the most convincing evidence of parapatric speciation comes in two forms. This is described by the following criteria:

- *Species populations that join, forming an ecotone can be interpreted as convincingly forming in parapatry if*:

    o   No evidence exists for a period of geographic separation between two closely related species

    o   Different loci are not in agreement along the cline

    o   Phylogenies including sister groups support different divergence times

- *An endemic species that exists within a specialized habitat next to its sister species that does not reside in the specialized habitat strongly suggests parapatric speciation.*

This has been exemplified by the grass species *Agrostis tenuis* that grows on soil contaminated with high levels of copper leeched from an unused mine. Adjacent is the non-contaminated soil. The populations are evolving reproductive isolation due to differences in flowering. The same phenomenon has been found in *Anthoxanthum odoratum* in lead and zinc contaminated soils.

Clines are often cited as evidence of parapatric speciation and numerous examples have been documented to exist in nature; many of which contain hybrid zones. These clinal patterns, however, can also often be explained by allopatric speciation followed by a period of secondary contact—causing difficulty for researchers attempting to determine their origin. Thomas B. Smith and colleagues posit that large ecotones are "centers for speciation" (implying parapatric speciation) and are involved in the production of biodiversity in tropical rainforests. They cite patterns of morphologic and genetic divergence of the passerine species *Andropa-*

dus virens. Jiggins and Mallet surveyed a range of literature documenting every phase of parapatric speciation in nature positing that it is both *possible* and *likely*. A study of tropical cave snails (*Georissa saulae*) found that cave-dwelling population descended from the above-ground population, likely speciating in parapatry.

In the Tennessee cave salamander, timing of migration was used to infer the differences in gene flow between cave-dwelling and surface-dwelling continuous populations. Concentrated gene flow and mean migration time results inferred a heterogenetic distribution and continuous parapatric speciation between populations.

Researchers studying *Ephedra*, a genus of gymnosperms in North American, found evidence of parapatric niche divergence for the sister species pairs *E. californica* and *E. trifurca.*

One study of Caucasian rock lizards suggested that habitat differences may be more important in the development of reproductive isolation than isolation time. *Darevskia rudis*, *D. valentini* and *D. portschinskii* all hybridize with each other in their hybrid zone; however, hybridization is stronger between *D. portschinskii* and *D. rudis*, which separated earlier but live in similar habitats than between *D. valentini* and two other species, which separated later but live in climatically different habitats.

Marine Organisms

It is widely thought that parapatric speciation is far more common in oceanic species due to the low probability of the presence of full geographic barriers (required in allopatry). Numerous studies conducted have documented parapatric speciation in marine organisms. Bernd Kramer and colleagues found evidence of parapatric speciation in Mormyrid fish (*Pollimyrus castelnaui*); whereas Rocha and Bowen contend that parapatric speciation is the primary mode among coral-reef fish. Evidence for a clinal model of parapatric speciation was found to occur in Salpidae. Nancy Knowlton found numerous examples of parapatry in a large survey of marine organisms.

Sympatric Speciation

Sympatric speciation is the process through which new species evolve from a single ancestral species while inhabiting the same geographic region. In evolutionary biology and biogeography, sympatric and sympatry are terms referring to organisms whose ranges overlap or are even identical, so that they occur together at least in some places. If these organisms are closely related (e.g. sister species), such a distribution may be the result of sympatric speciation. Etymologically, sympatry is derived from the Greek roots συν ("together") and πατρίς ("homeland"). The term was invented by Poulton in 1904, who explains the derivation.

Sympatric speciation is one of three traditional geographic categories for the phenomenon of speciation. Allopatric speciation is the evolution of species caused by the geographic isolation of two or more populations of a species. In this case, divergence is facilitated by the absence of gene flow. Parapatric speciation is the evolution of geographically adjacent populations into distinct species. In this case, divergence occurs despite limited interbreeding where the two diverging groups come into contact. In sympatric speciation, there is no geographic constraint to interbreeding. These categories are special cases of a continuum from zero (sympatric) to complete (allopatric) spatial segregation of diverging groups.

In multicellular eukaryotic organisms, sympatric speciation is a plausible process that is known to occur, but the frequency with which it occurs is not known. In bacteria, however, the analogous process (defined as "the origin of new bacterial species that occupy definable ecological niches") might be more common because bacteria are less constrained by the homogenizing effects of sexual reproduction and are prone to comparatively dramatic and rapid genetic change through horizontal gene transfer.

Evidence

Sympatric speciation events are quite common in plants, which are prone to acquiring multiple homologous sets of chromosomes, resulting in polyploidy. The polyploid offspring occupy the same environment as the parent plants (hence sympatry), but are reproductively isolated.

A number of models have been proposed for alternative modes of sympatric speciation. The most popular, which invokes the disruptive selection model, was first put forward by John Maynard Smith in 1966. Maynard Smith suggested that homozygous individuals may, under particular environmental conditions, have a greater fitness than those with alleles heterozygous for a certain trait. Under the mechanism of natural selection, therefore, homozygosity would be favoured over heterozygosity, eventually leading to speciation. Sympatric divergence could also result from the sexual conflict.

Disruption may also occur in multiple-gene traits. The medium ground finch (*Geospiza fortis*) is showing gene pool divergence in a population on Santa Cruz Island. Beak morphology conforms to two different size ideals, while intermediate individuals are selected against. Some characteristics (termed magic traits) such as beak morphology may drive speciation because they also affect mating signals. In this case, different beak phenotypes may result in different bird calls, providing a barrier to exchange between the gene pools.

A somewhat analogous system has been reported in horseshoe bats, in which echolocation call frequency appears to be a magic trait. In these bats, the constant frequency component of the call not only determines prey size but may also function in aspects of social communication. Work from one species, the large-eared horseshoe bat (*Rhinolophus philippinensis*), shows that abrupt changes in call frequency among sympatric morphs is correlated with reproductive isolation. A further well-studied circumstance of sympatric speciation is when insects feed on more than one species of host plant. In this case insects become specialized as they struggle to overcome the various plants' defense mechanisms. (Drès and Mallet, 2002)

Rhagoletis pomonella, the apple maggot, may be currently undergoing sympatric or, more precisely, heteropatric speciation. The apple feeding race of this species appears to have spontaneously emerged from the hawthorn feeding race in the 1800–1850 AD time frame, after apples were first introduced into North America. The apple feeding race does not now normally feed on hawthorns, and the hawthorn feeding race does not now normally feed on apples. This may be an early step towards the emergence of a new species. Some parasitic ants may have evolved via sympatric speciation. Isolated and relatively homogeneous habitats such as crater lakes and islands are among the best geographical settings in which to demonstrate sympatric speciation. For example, Nicaragua crater lake cichlid fishes include nine described species and dozens of undescribed species that have evolved by sympatric speciation.

Monostroma latissimum, a marine green algae, also shows sympatric speciation in southwest Japanese islands. Although panmictic, the molecular phylogenetics using nuclear introns revealed staggering diversification of population.

African cichlids also offer some evidence for sympatric speciation. They show a large amount of diversity in the African Great Lakes. Many studies point to sexual selection as a way of maintaining reproductive isolation. Female choice with regards to male coloration is one of the more studied modes of sexual selection in African cichlids. Female choice is present in cichlids because the female does much of the work in raising the offspring, while the male has little energy input in the offspring. She exerts sensory bias when picking males by choosing those that have colors similar to her or those that are the most colorful. This helps maintain sympatric speciation within the lakes. Cichlids also use acoustic reproductive communication. The male cichlid quivers as a ritualistic display for the female which produces a certain number of pulses and pulse period. Female choice for good genes and sensory bias is one of the deciding factors in this case, selecting for calls that are within her species and that give the best fitness advantage to increase the survivability of the offspring. Male-male competition is a form of intrasexual selection and also has an effect on speciation in African cichlids. Ritualistic fighting among males establishes which males are going to be more successful in mating. This is important in sympatric speciation because species with similar males may be competing for the same females. There may be a fitness advantage for one phenotype that could allow one species to invade another. Studies show this effect in species that are genetically similar, have the capability to interbreed, and show phenotypic color variation. Ecological character displacement is another means for sympatric speciation. Within each lake there are different niches that a species could occupy. For example, different diets and depth of the water could help to maintain isolation between species in the same lake.

Allochrony offers some empirical evidence that sympatric speciation has taken place, as many examples exist of recently diverged (sister taxa) allochronic species.

A rare example of sympatric speciation in animals is the divergence of "resident" and "transient" orca forms in the northeast Pacific. Resident and transient orcas inhabit the same waters, but avoid each other and do not interbreed. The two forms hunt different prey species and have different diets, vocal behaviour, and social structures. Some divergences between species could also result from contrasts in microhabitats. A population bottleneck occurred around 200,000 years ago greatly reducing the population size at the time as well as the variance of genes which allowed several ecotypes to emerge afterwards.

The European polecat (*Mustela putorius*) exhibited a rare dark phenotype similar to the European mink (*Mustela lutreola*) phenotype, which is directly influenced by peculiarities of forest brooks.

Controversy

For some time it was difficult to prove that sympatric speciation was possible, because it was impossible to observe it happening. It was believed by many, and championed by Ernst Mayr, that the theory of evolution by natural selection could not explain how two species could emerge from one if the subspecies were able to interbreed. Since Mayr's heyday in the 1940s and 50s, mechanisms have been proposed that explain how speciation might occur in the face of interbreeding, also known as gene flow. And even more recently concrete examples of sympatric divergence have been empirically studied. The debate now turns to how often sympatric speciation may actually occur in nature and how much of life's diversity it may be responsible for.

History

Ernst Mayr, a renowned 20th century evolutionary biologist from Germany argued that speciation cannot occur without geographic and thus reproductive, isolation. He stated that gene flow is the inevitable result of sympatry, which is known to squelch genetic differentiation between populations. Thus, a physical barrier must be present, he believed, at least temporarily, in order for a new biological species to arise. This hypothesis is the source of much controversy around the possibility of sympatric speciation. Mayr's hypothesis was popular and consequently quite influential, but is now widely disputed.

The first to propose what is now the most pervasive hypothesis on how sympatric speciation may occur was John Maynard-Smith. He came up with the idea of disruptive selection. He figured that if two ecological niches are occupied by a single species, diverging selection between the two niches could eventually cause reproductive isolation. By adapting to have the highest possible fitness in the distinct niches, two species may emerge from one even if they remain in the same area, and even if they are mating randomly.

Defining Sympatry

Investigating the possibility of sympatric speciation requires a definition thereof, especially in the 21st century, when mathematical modeling is used to investigate or to predict evolutionary phenomena. Much of the controversy concerning sympatric speciation may lie solely on an argument over what sympatric divergence actually is. The use of different definitions by researchers is, sadly, a great impediment to empirical progress on the matter. The dichotomy between sympatric and allopatric speciation is no longer accepted by the scientific community. It is more useful to think of a continuum, on which there are limitless levels of geographic and reproductive overlap between species. On one extreme is allopatry, in which the overlap is zero (no gene flow), and on the other extreme is sympatry, in which the ranges overlap completely (maximal gene flow).

The varying definitions of sympatric speciation fall generally into two categories: definitions based on biogeography, or on population genetics. As a strictly geographical concept, sympatric speciation is defined as one species diverging into two while the ranges of both nascent species overlap entirely – this definition is not specific enough about the original population to be useful in modeling.

Definitions based on population genetics are not necessarily spatial or geographical in nature, and can sometimes be more restrictive. These definitions deal with the demographics of a population, including allele frequencies, selection, population size, the probability of gene flow based on sex ratio, life cycles, etc. The main discrepancy between the two types of definitions tends to be the necessity for "panmixia". Population genetics definitions of sympatry require that mating be dispersed randomly – or that it be equally likely for an individual to mate with either subspecies, in one area as another, or on a new host as a nascent one: this is also known as panmixia. Population genetics definitions, also known as non-spatial definitions, thus require the real possibility for random mating, and do not always agree with spatial definitions on what is and what is not sympatry.

For example, micro-allopatry, also known as macro-sympatry, is a condition where there are two populations whose ranges overlap completely, but contact between the species is prevented because they occupy completely different ecological niches (such as diurnal vs. nocturnal). This can often be caused by host-specific parasitism, which causes dispersal to look like a mosaic across the landscape. Micro-allopatry is included as sympatry according to spatial definitions, but, as it does not satisfy panmixia, it is not considered sympatry according to population genetics definitions.

Mallet et al. (2002) claims that the new non-spatial definition is lacking in an ability to settle the debate about whether sympatric speciation regularly occurs in nature. They suggest using a spatial definition, but one that includes the role of dispersal, also known as cruising range, so as to represent more accurately the possibility for gene flow. They assert that this definition should be useful in modeling. They also state that under this definition, sympatric speciation seems plausible.

Current State of the Controversy

Evolutionary theory as well as mathematical models have predicted some plausible mechanisms for the divergence of species without a physical barrier. In addition there have now been several studies that have identified speciation that has occurred, or is occurring with gene flow. Molecular studies have been able to show that, in some cases where there is no chance for allopatry, species continue to diverge. One such example is a pair of species of isolated desert palms. Two distinct, but closely related species exist on the same island, but they occupy two distinct soil types found on the island, each with a drastically different pH balance. Because they are palms they send pollen through the air they could freely interbreed, except that speciation has already occurred, so that they do not produce viable hybrids. This is hard evidence for the fact that, in at least some cases, fully sympatric species really do experience diverging selection due to competition, in this case for a spot in the soil.

This, and the other few concrete examples that have been found, are just that; they're few, so they tell us little about how often sympatry actually results in speciation in a more typical context. The burden now lies on providing evidence for sympatric divergence occurring in non-isolated habitats. It is not known how much of the earth's diversity it could be responsible for. Some still say that panmixia should slow divergence, and thus sympatric speciation should be possible but rare (1). Meanwhile, others claim that much of the earth's diversity could be due to speciation without geographic isolation. The difficulty in supporting a sympatric speciation hypothesis has always been that an allopatric scenario could always be invented, and those can be hard to rule out – but with modern molecular genetic techniques can be used to support the theory.

In 2015 Cichlid fish from a tiny volcanic crater lake in Africa were observed in the act of sympatric speciation using DNA sequencing methods. A study found a complex combination of ecological separation and mate-choice preference had allowed two ecomorphs to genetically separate even in the presence of some genetic exchange.

Ecological Speciation

Ecological speciation is the process by which ecologically based divergent selection between different environments leads to the creation of reproductive barriers between

populations. This is often the result of selection over traits which are genetically correlated to reproductive isolation, thus speciation occurs as a by-product of adaptive
divergence.

European Holly (*Ilex aquifolium*). The genus *Ilex* is an example of ecological speciation.

Gasterosteus aculeatus, a documented case of ecological speciation

Ecological selection is "the interaction of individuals with their environment during
resource acquisition". Natural selection is inherently involved in the process of speciation, whereby, "under ecological speciation, populations in different environments,
or populations exploiting different resources, experience contrasting natural selection
pressures on the traits that directly or indirectly bring about the evolution of reproductive isolation". Evidence for the role ecology plays in the process of speciation exists.
Studies of stickleback populations support ecologically-linked speciation arising as a
by-product, alongside numerous studies of parallel speciation—of which, substantiates
speciation's occurrence in nature.

The key difference between ecological speciation and other kinds of speciation, is that
it is triggered by divergent natural selection among different habitats; as opposed to
other kinds of speciation processes, like random genetic drift, the fixation of incompatible mutations in populations experiencing similar selective pressures, or various
forms of sexual selection not involving selection on ecologically relevant traits. Ecological speciation can occur either in allopatry, sympatry, or parapatry. The only require-

ment being that speciation occurs as a result of adaptation to different ecological or micro-ecological conditions.

Centaurea solstitialis, a candidate species for incipient ecological speciation

Some debate exists over the framework concerning the delineation of whether a speciation event is ecological or nonecological. "The pervasive effect of selection suggests that adaptive evolution and speciation are inseparable, casting doubt on whether speciation is ever nonecological".

Parallel Speciation

Parallel speciation is where "greater reproductive isolation repeatedly evolves between independent populations adapting to contrasting environments than between independent populations adapting to similar environments". It is established that ecological speciation occurs and with much of the evidence, "...accumulated from top-down studies of adaptation and reproductive isolation".

Research and Supporting Evidence

Known examples of ecological speciation include three-spined stickleback fishes, distinct species of which emerged as the result to adaptation of different conditions along water depth clines in freshwater lakes. Ancestors of the genus *Ilex* (holly) became isolated from the remaining *Ilex* when the Earth mass broke away into Gondwana and Laurasia about 82 million years ago, resulting in a physical separation of the groups (allopatry) and beginning a process of change to adapt to new conditions; over time survivor species of the holly genus adapted to different ecological niches. The invasive weed species *Centaurea solstitialis* is a candidate to be a case of incipient ecological speciation; in less than 200 years, incipient reproductive isolation appeared as a result to adaptation to different ecological conditions between native and non-native ranges.

Mosquito Fish

Parallel speciation occurs for example in mosquito fish in the Bahamas, where

Gambusia fish inhabit "blue holes"—carbonate caves and depressions flooded with water throughout the islands. Some of the holes contain the piscivorous predator fish *Gobiomorus dormitor*, while others have no major predators (excluding birds). The authors of the study tested for ecological speciation by measuring three different data sets: morphological data (to test for divergent natural selection), molecular data (to test for "replicated trait evolution in independent populations" with similar phenotypes), and mate-choice trials (to test for reproductive isolation between "ecologically divergent pairs of populations than ecologically similar ones": a by-product resulting from divergent traits). The study allowed for a natural experiment to test the effects of predator-mediated natural selection and its by-product: ecological speciation. The results suggested a "strong confirmation of the ecological speciation hypothesis" and amply supported parallel speciation taking place within the different blue holes.

Skinks

Another example is in skinks, where the *Plestiodon* genus (formerly *Eumeces*) has a complex evolutionary history with ecological speciation and parallel speciation of the three species (within two morphotypes): *E. skiltonianus* and *E. lagunensis*, and *E. gilberti*. *E. gilberti* occupies dry, low-elevation habitats, has a larger body size, and uniform, solid color scales. The other two species inhabit higher-elevation regions, have smaller bodies, and exhibit colored stripes. The members of the group have similar phenotypic stages during early development but differ in their morphology in later stages. A phylogenetic analysis using mtDNA of the entire group, (including all species and subspecies of the *Eumeces* group inhabiting the western United States) combined with comparative approaches to morphology, and geographic distribution showed "instances of parallel morphological evolution...and provide evidence that this system is consistent with a model of ecological speciation" due to "the similarity in early ontogenetic trajectories and the close association between differences in body size and color pattern[s]" of each morphotype.

References

- Koeslag, Johan H. (December 21, 1995). "On the Engine of Speciation". Journal of Theoretical Biology. Amsterdam, the Netherlands: Elsevier. 177 (4): 401–409. ISSN 0022-5193. doi:10.1006/jtbi.1995.0256

- Ridley, Mark. "Speciation - What is the role of reinforcement in speciation?". Retrieved 2015-09-07. Adapted from Evolution (2004), 3rd edition (Malden, MA: Blackwell Publishing), ISBN 978-1-4051-0345-9

- Liebers, Dorit; Knijff, Peter de; Helbig, Andreas J. (2004). "The herring gull complex is not a ring species". Proc Biol Sci. 271 (1542): 893–901. PMC 1691675. PMID 15255043. doi:10.1098/rspb.2004.2679

- Anders Ödeen and Ann-Britt Florin (2002), "Sexual selection and peripatric speciation: the Kaneshiro model revisited", Journal of Evolutionary Biology, 15: 301–306

- Jerry A. Coyne and H. Allen Orr (2004), Speciation, Sinauer Associates, pp. 111–124, ISBN 0-87893-091-4

- Sherwood, Jonathan (September 8, 2006). "Genetic Surprise Confirms Neglected 70-Year-Old Evolutionary Theory" (Press release). University of Rochester. Retrieved 2015-09-10

- Nathalie Seddon and Joseph A. Tobias (2007), "Song divergence at the edge of Amazonia: an empirical test of the peripatric speciation model", Biological Journal of the Linnean Society, 90: 173–188

- Martens, Koen (May 1997). "Speciation in ancient lakes". Trends in Ecology & Evolution. 12 (5): 177–182. doi:10.1016/S0169-5347(97)01039-2

- Pinker, Steven (June 18, 2012). "The False Allure of Group Selection". edge.org. Edge Foundation, Inc. Retrieved 2015-09-15

- Michael Doebeli and Ulf Dieckmann (2003), "Speciation along environmental gradients", Nature, 421: 259–264, doi:10.1038/nature01274

- Rosemary G. Gillespie (2005), "Geographical context of speciation in a radiation of Hawaiian Tetragnatha spiders (Aranae, Tetragnathidae", The Journal of Arachnology, 33: 313–322

- Rabeling, Christian; Schultz, Ted R.; Pierce, Naomi E.; Bacci Jr, Maurício (August 2014). "A Social Parasite Evolved Reproductive Isolation from Its Fungus-Growing Ant Host in Sympatry". Current Biology. 24: 2047–2052. doi:10.1016/j.cub.2014.07.048. Retrieved 22 August 2014

- Maynard Smith, John (March 14, 1964). "Group Selection and Kin Selection". Nature. London: Nature Publishing Group. 201 (4924): 1145–1147. ISSN 0028-0836. doi:10.1038/2011145a0

- Jessica E. Garb (1999), "An Adaptive Radiation of Hawaiian Thomisidae: Biogreographic and Genetic Evidence", The Journal of Arachnology, 27: 71–78

- Mayr, E. (1942). Systematics and the origin of species from the viewpoint of a zoologist. New York: Columbia University Press. p. 337. ISBN 0-674-86250-3

- Mallet, J.; Meyer, A.; Nosil, P.; Feder, J. L. (2009). "Space, sympatry and speciation". Journal of Evolutionary Biology. 22 (11): 2332–41. PMID 19732264. doi:10.1111/j.1420-9101.2009.01816.x

- Lucinda P. Lawson; et al. (2015), "Divergence at the edges: peripatric isolation in the montane spiny throated reed frog complex", BMC Evolutionary Biology, 15 (128), doi:10.1186/s12862-015-0384-3

Permissions

All chapters in this book are published with permission under the Creative Commons Attribution Share Alike License or equivalent. Every chapter published in this book has been scrutinized by our experts. Their significance has been extensively debated. The topics covered herein carry significant information for a comprehensive understanding. They may even be implemented as practical applications or may be referred to as a beginning point for further studies.

We would like to thank the editorial team for lending their expertise to make the book truly unique. They have played a crucial role in the development of this book. Without their invaluable contributions this book wouldn't have been possible. They have made vital efforts to compile up to date information on the varied aspects of this subject to make this book a valuable addition to the collection of many professionals and students.

This book was conceptualized with the vision of imparting up-to-date and integrated information in this field. To ensure the same, a matchless editorial board was set up. Every individual on the board went through rigorous rounds of assessment to prove their worth. After which they invested a large part of their time researching and compiling the most relevant data for our readers.

The editorial board has been involved in producing this book since its inception. They have spent rigorous hours researching and exploring the diverse topics which have resulted in the successful publishing of this book. They have passed on their knowledge of decades through this book. To expedite this challenging task, the publisher supported the team at every step. A small team of assistant editors was also appointed to further simplify the editing procedure and attain best results for the readers.

Apart from the editorial board, the designing team has also invested a significant amount of their time in understanding the subject and creating the most relevant covers. They scrutinized every image to scout for the most suitable representation of the subject and create an appropriate cover for the book.

The publishing team has been an ardent support to the editorial, designing and production team. Their endless efforts to recruit the best for this project, has resulted in the accomplishment of this book. They are a veteran in the field of academics and their pool of knowledge is as vast as their experience in printing. Their expertise and guidance has proved useful at every step. Their uncompromising quality standards have made this book an exceptional effort. Their encouragement from time to time has been an inspiration for everyone.

The publisher and the editorial board hope that this book will prove to be a valuable piece of knowledge for students, practitioners and scholars across the globe.

Index